This book provides a panoramic view during 1927–38 of the development of a physical theory that has been on the cutting edge of theoretical physics ever since P. A. M. Dirac's quantization of the electromagnetic field in 1927: quantum electrodynamics. Like the classic papers chosen for this volume, the introductory Frame-setting essay emphasizes conceptual transformations that carried physicists to the threshold of renormalization theory. The published papers and correspondence of Bohr, Dirac, Heisenberg and Pauli take us on a fascinating analysis into the meaning and structure of a scientific theory. The subject matter of this book goes beyond the historical and philosophical into current physics. Unavailability of English-language versions of certain key papers, some of which are provided in this book, has prevented their implications from being fully realized. Awareness of research from sixty years ago could well provide insights for future developments.

Early quantum electrodynamics:
a source book

Early quantum electrodynamics:

a source book

ARTHUR I. MILLER
Department of History, Philosophy and Communication of Science
University College London

Translations from the German by WALTER GRANT

CAMBRIDGE
UNIVERSITY PRESS

Published by the Press Syndicate of the University of Cambridge
The Pitt Building, Trumpington Street, Cambridge CB2 1RP
40 West 20th Street, New York, NY 10011–4211, USA
10 Stamford Road, Oakleigh, Melbourne 3166, Australia

First published 1994
First paperback edition 1995

Printed in Great Britain at the University Press, Cambridge

A catalogue record for this book is available from the British Library

Library of Congress cataloguing in publication data

Miller, Arthur I.
Early quantum electrodynamics: a source book / Arthur I. Miller;
translations from the German by Walter Grant.
p. cm.
Includes bibliographical references.
ISBN 0-521-43169-7
1. Quantum electrodynamics – History. I. Title.
QC680.M55 1994
537.6 – dc20 93–12774 CIP

ISBN 0 521 43169 7 hardback
ISBN 0 521 568919 paperback

KT

For
Norma

Contents

II Selected papers

Preface

Thus far, studies in the history of twentieth-century physics have focused on developments until 1927. Testimony to the vigor of this work is the rich secondary literature on the special and general theories of relativity and on quantum mechanics into its interpretive phase.

Historical research into the genesis of quantum electrodynamics is only just beginning.[1] There are good reasons for this hiatus, chief among them being the complexity of the subject matter. The goal of this book is to provide a properly introduced corpus of primary source materials in English to physics researchers and students whose day-to-day activities preclude literature searches, and to historians and philosophers of science interested in the genesis of a theory that has been on the cutting edge of physics ever since P. A. M. Dirac's quantization of the radiation field in 1927: quantum electrodynamics.

Like the papers chosen for this volume, the Frame-setting essay emphasizes conceptual transformations during 1927–38, which carried physicists to the threshold of renormalization theory.[2] For the most part the leaders in fundamental developments in quantum electrodynamics were the same physicists whose focus on conceptual matters led to the fully interpreted quantum mechanics in 1927: Niels Bohr, P. A. M. Dirac, Werner Heisenberg and Wolfgang Pauli. This constitutes the subject matter of Chapter 1 of the Frame-setting essay. Throughout Chapter 1 runs the metaphor of the harmonic oscillator representation for a bound electron, which found its highest development in the second-quantization methods discussed in Chapter 2.

After 1927 the overall strategy was to extend quantum theory into the domain of high energies and correspondingly small spatial distances by means of appropriate correspondence-limit procedures. The new framework would be interpreted in an intuitive or '*anschaulich*' manner, according to the new meaning of this concept that had emerged from the quantum mechanics. This point is the Ariadne's thread that runs throughout the Frame-setting essay.

Chapter 3 focuses on Dirac's 1928 relativistic theory of the electron, various interpretations of the negative energy states, the formal invention of quantum electrodynamics in 1929 by Heisenberg and Pauli, and Dirac's postulation in 1931 of the positron. With this background, Chapter 4 sets the classic papers collected in this volume into their proper historical context, with analysis of the search for methods to calculate a finite electron mass, to understand second quantization and to seek a connection between spin and statistics. Each paper is analyzed in conjunction with relevant correspondence and other pertinent developments.

Chapter 5 focuses on conceptual developments in the theory of nuclear forces that affected quantum electrodynamics. The question arises of why descriptions of nuclear forces carried by particles did not become depictions which might have aided invention of renormalization in the 1930s.

Instead of reprinting a great many papers in part, I have chosen to reprint a few in their entirety. My selection criteria were threefold – economics, language and theme. In order to produce a book whose price was not astronomical, some difficult choices had to be made. Since the book is meant primarily for an English-speaking audience, then (with one exception) only foreign-language papers were considered which would then be translated. A further restriction was that, to the best of my knowledge, the chosen papers were not translated elsewhere. Most of the relevant English-language papers from the 1930s are in Julian Schwinger's reprint volume *Selected Papers on Quantum Electrodynamics*, which contains the post-World War II classic papers on renormalization theory.

My selection of papers is based ultimately on the overarching problematics of quantum electrodynamics in the 1930s and its projected reformulation as a quantized theory of wave fields, that is, a quantum field theory, namely to formulate a version of quantum theory consistent with special relativity and gauge invariance that could treat with no divergences the interaction between light and electrons, *and* was extendable to the nuclear force.

In order to accomplish this feat, there were two not unconnected lines of research. During 1928–33 covariant methods consistent with gauge invariance were sought to quantize the electromagnetic field and to formulate the basis of a quantum electrodynamics. From this work emerged a divergent self-energy for the electron that could not be dealt with as cavalierly as the one from classical electromagnetic theory. Efforts to rid quantum electrodynamics of this infinite quantity commenced seriously in the line of development begun in 1933 when Dirac invented a procedure honed to perfection in 1934 by Heisenberg for subtracting infinite terms occurring, for example, in the vacuum expectation value for the charge density. Whilst persistence of the electron's infinite self-energy caused Pauli's disenchantment with quantum electrodynamics, Heisenberg continued to spin bold techniques to dispose of this divergent quantity. This drama emerges from the Heisenberg–Pauli correspondence, which plays a central part in the Frame-setting essay. Just like the great theories that were its ancestors, for Dirac, Heisenberg and Pauli the fully developed quantum electrodynamics would possess no infinite quantities.

In the 1930s there were instances where Heisenberg and Pauli suggested that adherence to correspondence-limit procedures was the source of problems in quantum electrodynamics, such as the divergent self-energy of the electron, for example, taking classical Hamiltonian methods as the starting point for calculations. Heisenberg suggested alternative schemes that failed. Then, in 1943, he formulated the S-matrix theory in a manner analogous to the way in which he invented quantum mechanics in 1925: Heisenberg sought to construct a version of quantum electrodynamics based only on measurable quantities. He expected this theory to provide clues about how to deal with phenomena occurring in spatial regions smaller than a 'fundamental length'.

It turned out that no fundamental length was needed and that all the technical apparatus and basic concepts for a renormalized quantum electrodynamics were already in place in the 1930s. As Freeman J. Dyson wrote in (1949b) about the new quantum electrodynamics in which Heisenberg's S-matrix plays a central role, when properly reformulated: '[U]sing no new ideas or techniques, one arrives at an S matrix from which the well-known divergences seem to have conspired to eliminate themselves.'

How did it come about that assembling already existing methods and concepts into a renormalized quantum electrodynamics eluded physicists in the 1930s? This is one among several historical–philosophical–physical problems central to exploring the genesis of quantum electrodynamics.[3] It emerges from the primary papers collected here that exploring this problem depends on properly understanding issues critical to physicists in the 1930s, such as: Are electromagnetic fields measurable in relativistic quantum theory? Why are there two methods for quantization of wave fields, commutators and anticommutators? Is there a relation between second quantization and the appearance in a second-quantized version of quantum electrodynamics of divergent quantities, such as the electron's self-energy,[4] the photon's self-energy and vacuum polarization?

Taking into account the existence in the 1930s of proper mathematical methods, could pursuance of correspondence-limit procedures coupled with the quest to further redefine the concept of intuition have led to a renormalized quantum electrodynamics? This scenario is less dependent on empirical data than on the standard history in which the Lamb shift is pivotal (for example, Pais, 1986, and Weinberg, 1977). We have two alternative scenarios for progress in quantum electrodynamics: one emphasizing rationalism coupled with conceptual analysis; the other emphasizing empiricism. The first alternative *emerges* from the primary papers and their historical analysis in this volume.

Except for one selection, all papers have been translated from their original German and one from French. The German translations were rendered by Dr Walter Grant and edited by me. I translated from the French Dirac's 1933 lecture at Solvay and the discussion session. The English-language paper by Dirac, 'Discussion of the infinite distribution of electrons in the theory of the positron', is essential for understanding Heisenberg's 1934 paper 'Remarks on the Dirac theory of the positron'.

A. I. M.
London

Notes to the Preface

1 See, for example, Weinberg (1977), Brown (1981, 1991), Kragh (1981), Miller (1984, 1985, 1990, pp. 139–52), Darrigol (1986, 1988a, b), Pais (1986) and Schweber (1986a). For further references to the second literature see Hovis and Kragh (1991).

2 A useful internalist history of quantum electrodynamics is Pais (1986), which has extensive references to the primary literature.

3 For a sample of recent philosophical writings on quantum field theory, see Harré (1986) and Brown and Harré (1988).

4 In a properly covariant quantum electrodynamics, the self-energy of the photon is exactly zero. See, for example, Schweber (1961) and Sakurai (1967).

Acknowledgements

For travel and research funds essential for this book it is a pleasure to acknowledge support from the Section for History and Philosophy of Science of the National Science Foundation, the National Endowment for the Humanities, the Fritz Thyssen Stiftung, and the University Professor Fund of the University of Lowell, Lowell, MA. This book originally stood under the aegis of Prof. Edward H. Madden's *Source Book* program. I thank Prof. Madden as well as Prof. Gerald Holton for encouragement in this period. It is a pleasure to acknowledge conversations with Profs Steven Weinberg and Viktor F. Weisskopf which took place during the early stages of research on this book. For their helpful comments on the manuscript I am grateful to Dr Rom Harré and Prof. Michael Redhead. Needless to say, I alone am responsible for the choice of papers included, as well as for the line of development in the Frame-setting essay.

Notes to the Reader

I use the author–date system of references, so Heisenberg (1926a) means the paper listed in the References under Heisenberg and dated 1926a. Sometimes I have placed the reference date in the written text to improve the readability. Thus 'In (1918) Bohr proposed . . .' cites the reference Bohr (1918).

Primary papers collected in this volume are referred to throughout the text as, for example, [1], which means paper number 1 in the Contents, which is Heisenberg's 'The self-energy of the electron'.

Equations from papers listed in the Contents are referred to as Eq. (1)–[6], meaning Eq. (1) in Heisenberg's paper 'Remarks on the Dirac theory of the positron'.

Quotations from *AHQP* (*Archive for History of Quantum Physics*) are taken from interviews on deposit at the American Institute of Physics, New York City, the American Philosophical Society, Philadelphia, the University of California, Berkeley, and the Niels Bohr Institute, Copenhagen.

I *Frame-setting essay*

1 From quantum mechanics toward quantum electrodynamics*

Prior to and into the first decade of Niels Bohr's 1913 atomic theory, physicists dealt with physical systems in which the usual space and time pictures of classical physics were assumed trustworthy and so could be extrapolated to any sort of matter in motion, for example, electrons move like billiard balls and light behaves analogously to water waves. In the German scientific milieu this visual imagery was accorded a reality status higher than viewing merely with the senses and was referred to as 'ordinary intuition [*gewöhnliche Anschauung*]'. Ordinary intuition is the visual imagery abstracted from phenomena that we have actually witnessed in the world of sense perceptions. The concept of ordinary intuition was much debated during 1923–27 by physicists like Niels Bohr, Werner Heisenberg, Wolfgang Pauli and Erwin Schrödinger, all of whom used this term with proper Kantian overtones, and all of whom lamented its loss in the new quantum mechanics.

Ordinary intuition is associated with the strong causality of classical mechanics. According to the law of causality, initial conditions (position and velocity or momentum and energy) can be ascertained with in-principle perfect accuracy. Consequently any system's continuous development in space and time can be traced with in-principle perfect accuracy using Newton's laws of motion and conservation of energy and momentum. Any limitations to the accuracy of measurements are assumed not to be intrinsic to the phenomena, that is they are interpreted to be systematic measurement errors that ideally can be made to vanish. Classical physics offers the connection pictures–causality–conservation laws.

In the first decade of the twentieth century the consensus among physicists was that a method would be found to extend our intuition from classical physics into the atomic domain. They believed that laws governing the behavior of individual atoms would not be statistical. For example, Ernest Rutherford's law for how many of a large number of atoms undergo radioactive decay in a certain time period is a statistical law in the sense of classical physics, where statistics and probability are interpreted as reflecting our ignorance of the underlying dynamics of individual processes. Rather, some complex form of the causal Newtonian mechanics would in time be formulated for Rutherford's

* Among the secondary literature on the history of quantum mechanics with extensive bibliographies are: Jammer (1966, 1974), Hendry (1984) and Miller (1984). The interested reader should also consult the *Isis Critical Bibliography*.

model of the atom as a nucleus surrounded by electrons. (For example, see Poincaré, 1913.)

1.1 Niels Bohr's atomic theory, 1913–23

How important imagery was (and still is) to physicists is clear from Bohr's seminal papers of 1913 (Bohr, 1913). Despite his theory's violation of classical mechanics, Bohr emphasized that mathematical symbols from classical mechanics permitted visualization of the atom as a miniscule Copernican system. Although suitably quantized laws of classical mechanics are used to calculate the electron's allowed orbits, or stationary states, classical mechanics can neither depict nor describe the electron in transit. In transit the orbital electron behaves like the Cheshire cat because the quantum jump or 'essential discontinuity', is unvisualizable. In contrast, classical electrodynamics could not at all account for any characteristics of radiation emitted in the transition. In (1918) Bohr proposed a method to extend classical electrodynamics into the realm of the atom by means of what he would call in 1920 the 'correspondence principle'.[1]

1.2 The coupling mechanism

By 1923 the picture of a planetary atom was beginning to wither away. Besides its lack of success in dealing with atoms more complex than hydrogen, the problem of dispersion altered dramatically Bohr's atomic theory because the response of atomic electrons to incident light could not always be correlated with their simple motion in Keplerian orbits.

In (1923) Bohr proposed that 'fundamental difficulties' facing his theory all had their common denominator in the problem of the interaction of light with atoms. The key point was to reconcile essential discontinuities of atomic physics with the inherent continuity of classical electrodynamics. One approach suggested by Bohr involved the light quantum and maintaining energy and momentum conservation in individual processes. But this was an unsatisfactory solution because the 'picture [*Bild*] of light quanta precludes explaining interference'. This had been the principal criticism against the light quantum ever since its invention by Einstein in 1905. Yet the light quantum's undeniable usefulness for explaining certain phenomena reinforced Bohr's belief that a contradiction-free description of atomic processes could not be arrived at by 'use of conceptions borrowed from classical electrodynamics'. Since in classical physics the conservation laws are linked with a continuous space-time description then these laws may 'not possess unlimited validity'.

Bohr was not prepared to take this step at that time. Instead his guide in the atomic domain would be the correspondence principle, which would enable him 'to make assumptions that are quite foreign to the quantum theory'. One such assumption was the 'coupling mechanism', which has its roots in papers of Paul Ehrenfest (1906) and Peter Debye (1910) who discussed cavity radiation and Planck's law using the method of normal modes proposed by Rayleigh and Jeans. Quantizing the normal modes of the radiation field in integral multiples of $h\nu$, where ν is the frequency of a normal mode, they emphasized the formal analogy between the mathematical description of normal modes of a wave field and oscillators.

Bohr proposed that the coupling mechanism is activated when an atom is illuminated by light containing frequencies capable of inducing transitions between stationary states. The atom responds to radiation like a number of classical oscillators, each of which oscillates with the frequency of a quantum transition. The probability of an energy exchange can be calculated by combining the coupling principle with the correspondence principle; in this way Einstein's A coefficient for spontaneous transitions enters atomic physics. The coupling mechanism permitted Bohr to renounce the 'so-called hypothesis of light quanta'. Bohr mentioned that 'a line of thought of this kind [the coupling mechanism] was first followed out closely in a work by Rudolf Ladenburg'.

Ladenburg had worked for many years on both experimental and theoretical aspects of dispersion. In (1921), independently of Bohr's work, he decided to exploit the moderately successful part of the classical theory of dispersion in which an atom responds to incident radiation like a charged oscillator (see Eq. (B) in note 1).

In (1923) Ladenburg and Fritz Reiche proposed a mathematical formulation for the coupling mechanism. The classical electrodynamical treatment of dispersion yields for the polarizability α

$$\alpha = \frac{e^2}{4\pi m} \sum_k \frac{N_k}{\nu_{0k}^2 - \nu^2}, \tag{1.1}$$

where N_k is the number of dispersion electrons of type k (a constant fitted to data) with mass m and charge e, each assumed to behave like a classical oscillator of proper frequency ν_{0k}, ν is the frequency of the incident radiation, and the atom's dipole moment $P(t) = \alpha E(t)$, where $E(t)$ is the applied electric field. Using the correspondence principle, Eq. (1.1) is rewritten as

$$\alpha = \frac{e^2}{4\pi m} \sum_k \frac{a_{ki}}{\nu_{ik}^2 - \nu_0^2}, \tag{1.2}$$

where a_{ki} is related to the Einstein A coefficient for a spontaneous transition between stationary states i and k with emission of light of frequency ν_{ik}. Ladenburg and Reiche interpreted Eq. (1.2) 'formally' as the atom responding to radiation like '*Ersatzoszillatoren*', each one having the frequency of a possible atomic transition.

1.3 Virtual oscillators

In 1924 the coupling mechanism provided Bohr, Hendrik Kramers and John C. Slater a way to avoid interpreting the Compton effect in terms of light quanta (Bohr, Kramers and Slater, 1924). To Bohr the tension between the two conceptions of light would have to be resolved on the basis of the wave theory. For although there are essential discontinuities in atomic physics, our 'ordinary intuition [*Anschauung*]' requires light to be a wave phenomenon. In order to exclude light quanta, Bohr, Kramers and Slater resorted to combining the most extreme consequence of the first method of 1923 (renouncing energy conservation) with the oscillators in the coupling mechanism, which they referred to as 'virtual oscillators'. Besides emitting real radiation in spherical waves in response to incident radiation, the virtual oscillators were assumed to emit a field carrying only probability for inducing atomic transitions. The virtual radiation field of one atom could induce an upward atomic transition in another atom without the source atom undergoing the corresponding downward transition, thereby violating energy conservation and causality in individual processes. In this way Bohr, Kramers and Slater sought to reconcile discontinuous atomic transitions with the continuous radiation field.

Bohr, Kramers and Slater interpreted the Compton effect as follows: Each illuminated electron in the target crystal emits coherent secondary wavelets that can be understood as the usual sort of light scattered from a harmonic (here virtual) oscillator. As a consequence of the virtual radiation field, the scattered electron has a probability of having momenta in any direction. In this way the Compton effect can be understood as a continuous process. Bohr considered such a radical version of his theory necessary in order to avoid the paradoxical circumstance of having to deal with an entity that can be both wave and particle simultaneously.

Bohr, Kramers and Slater did not use the term 'picture [*Bild*]' of the atom to mean visualization. Is it not impossible to visualize an electron in a stationary state when the atomic electron is represented by as many oscillators as there are transitions to and from this state? Rather, they meant the term 'picture' to refer to the interpretation of the mathematical framework. The picture of the Copernican atom had been *imposed* on the 1913–23 Bohr theory owing to Bohr's use of the language (semantics) of 'ordinary mechanics' (Bohr, 1913). The 1924 Bohr, Kramers and Slater version of Bohr's theory started the movement toward defining the image of atomic theory to be *given* by its mathematical scheme. Moreover, Bohr, Kramers and Slater raised the concept of probability from a strictly mathematical entity to one that actually produced physical phenomena such as atomic transitions.

Heisenberg, among others, was much impressed by the 'intermediate kinds of reality' (*AHQP*: 13 February 1963) offered by the virtual oscillator representation. This situation augured to Heisenberg that 'cheap solutions would not be found' (*AHQP*: 13 February 1963). But whereas subsequent work on dispersion by Max Born, Heisenberg and Kramers used virtual oscillators, neither violations of energy nor of momentum conservation were well received.

Central to the virtual oscillator formalism is that the bound electron can be represented purely in terms of measurable quantities through the equation for the atom's dipole moment (see note 1)

$$P(t) = \sum_q U_q \exp\{2\pi i \nu_q t\}, \tag{1.3}$$

where the intensity of a spectral line is given by the magnitude squared of the amplitude U_q, and the line's measured frequency is ν_q. Eq. (1.3) became the basis of Kramers's work on dispersion begun in (1924) and concluded in December (1925) with the Kramers–Heisenberg paper, which turned out to be the high-water mark of the Bohr theory. No further progress was made.

1.4 Quantum mechanics versus wave mechanics

By mid-1925 fundamental conceptual problems focused on lack of visualization of atomic phenomena: Owing to the virtual oscillator representation, bound electrons had lost their localization and visualization; owing to Bose–Einstein statistics (as it was interpreted in 1926 – see Heisenberg, 1926b) free electrons had lost their distinguishability and individuality too; and then there was the wave–particle duality of light and matter. Not surprisingly, lack of visualizability entailed linguistic problems. For example, the defining equation for a light quantum is $E = h\nu$. The quantity E connotes localization, but ν is a 'radiation frequency defined by experiments on interference phenomena' (Bohr, Kramers and Slater, 1924).

Faced in 1925 with experimental refutation of the Bohr–Kramers–Slater theory, and the possibility that the light quantum might be real, Bohr reluctantly renounced 'intuitive pictures' of atomic processes, while accepting the conservation laws for individual atomic processes (Bohr, 1925).

Suffice it to say that the virtual oscillator representation was central to Heisenberg's invention of the new quantum mechanics or matrix mechanics in June 1925, based 'exclusively on relations between quantities which in principle are [empirically] observable' (Heisenberg, 1925).

Although renunciation of the picture of a bound electron had been a necessary prerequisite to Heisenberg's invention of the new quantum mechanics, the lack of an 'intuitive [*anschauliche*]' interpretation was of great concern to Bohr, Born and Heisenberg. This concern emerges from their scientific papers of the period 1925–27.

With the publication in early 1926 of Erwin Schrödinger's wave mechanics, the quest for some sort of visualization of atomic processes intensified and took a subjective turn in the published scientific literature. Schrödinger (1926a) wrote that he formulated the wave mechanics because he 'felt discouraged not to say repelled . . . by lack of visualizability [*Anschaulichkeit*]' of the quantum mechanics. He offered a visual representation based on the ordinary intuition of atomic processes occurring without discontinuities as wave phenomena.

To summarize: In mid-1926 there were two seemingly dissimilar atomic theories. Heisenberg's quantum mechanics was corpuscular based and yet renounced any visualization of the bound corpuscle itself. Its mathematical apparatus was unfamiliar to most physicists. Wave mechanics was a continuum theory based on matter as waves. Its familiar mathematical apparatus led to a calculational breakthrough, and its claim to restore customary intuition was welcomed by many physicists including Einstein.

Heisenberg thought otherwise. On 8 June 1926 he wrote to Pauli (1979): 'The more I reflect on the physical portion of Schrödinger's theory the more disgusting I find it . . . . What Schrödinger writes on the visualizability [*Anschaulichkeit*] of his theory . . . I consider trash. The great accomplishment of Schrödinger's theory is the calculation of matrix elements.'

The tension between the quantum and wave mechanics increased with the appearance of Born's quantum theory of scattering in mid-1926. To Born (1926) neither scattering problems nor transitions in atoms could be understood using quantum mechanics, which denied 'exact representation of processes in space and time', or wave mechanics, which denied visualization in phenomena with more than one particle. Problems concerning scattering and transitions required the 'construction of new concepts', and for his vehicle Born chose to use wave mechanics, which allows for at least the possibility of visualization.

1.5 Intrinsic symmetry

Schrödinger (1926a) expressly cautioned against treating many-body systems owing to '*Anschauungsfrage*' pertaining to the interpenetration of wave packets. Heisenberg responded to Schrödinger's wave mechanics with a remarkable paper of June 1926 entitled 'Many-body problem and resonance in quantum mechanics' (1926a), where he pushed aside Schrödinger's proposed ban on *Anschauungsfrage* and invented the totally nonclassical concept of intrinsic symmetry expressly to explore many-body systems. The concept of intrinsic symmetry forcefully revealed some of the deepest consequences of lack of continuity and visualization for atomic systems.

The puzzles facing Heisenberg were: (1) that the spacing between singlet and triplet systems in the spectrum of the helium atom was too large to be caused only by a spin interaction between two atomic electrons; (2) that at first sight in either the quantum or wave mechanics denumeration of the number of stationary states of a quantum mechanical many-electron system should be the same as the classical or Boltzmann statistics. Yet far fewer stationary states were seen to occur in nature. Why?

Heisenberg found these two problems to be connected to a much deeper conceptual puzzle, namely the relationship to quantum mechanics of Bose–Einstein statistics (one of whose consequences is the indistinguishability of particles); and to what Heisenberg referred in the 'many-body paper' as 'supplementary rules', such as Pauli's exclusion principle. These concepts had yet to find their place in either the quantum or wave mechanics.

Despite strenuous efforts by Born, Heisenberg and Pauli, the stationary states of the simplest three-body atomic systems, the H_2^+ ion and helium atom, could not be deduced from Bohr's atomic theory. Heisenberg later recalled (*AHQP*: 11 February 1963) these results of 1922 and 1923 as the 'first moment when really this confidence [in the Bohr theory] was shaken'. In 1926 Schrödinger's pessimism on conceptual difficulties in the many-body problem led Heisenberg back to the helium atom, but now armed with the new quantum mechanics.

In the 'many-body paper' Heisenberg motivated the helium atom calculation with two identical atomic systems a and b coupled through an interaction energy V symmetric in the two systems. The stationary states of the unperturbed systems are assumed to be nondegenerate but the term spectrum of the total unperturbed system

$$H_{nm} = H_n^a + H_m^b \tag{1.4}$$

is degenerate, that is $H_{nm} = H_{mn}$. The interaction V removes the energy degeneracy and the combined system pulsates or resonates like two coupled oscillators in two normal modes with energies that Heisenberg calculated with first-order perturbation theory as:

$$W = SVS^{-1}, \tag{1.5}$$

where W is a diagonal matrix with elements

$$W_1 = J + K \tag{1.6}$$

and

$$W_2 = J - K. \tag{1.7}$$

J is the matrix element for the interaction energy between states m and n. The degeneracy is removed by the matrix element K, which is the interaction energy when the systems exchange quantum states. The elements of the transformation matrix S are

$$S_{11} = S_{22} = \frac{1}{2^{1/2}} \quad \text{and} \quad S_{12} = -S_{21} = \frac{1}{2^{1/2}}. \tag{1.8}$$

Heisenberg demonstrated that the 'decisive result' was that matrix elements vanish for transitions between systems with energies W_1 and W_2 because operators for radiative transitions must be symmetric functions in the coordinates of both systems. The problem remains of whether in nature the term series corresponding to the eigenvalues W_1 and W_2 exist separately or in combination.

In order to make the resonance phenomenon more transparent, and to calculate matrix elements, Heisenberg turned to Schrödinger's wave functions (which were the 'great accomplishment' of wave mechanics because they facilitated 'calculation of matrix elements', as he wrote to Pauli on 8 June 1926, in Pauli, 1979). The spatial wave function of the unperturbed system is $\varphi_n^a\varphi_m^b$. Linear transformation with the matrix S yields for $W_1 = J + K$ the spatial wave function ψ_S that is symmetric under exchange of the two systems a and b

$$\psi_S = \frac{1}{2^{1/2}}\{\varphi_n^a\varphi_m^b + \varphi_n^b\varphi_m^a\}, \tag{1.9}$$

and for $W_2 = J - K$ the antisymmetric spatial wave function ψ_A, which is

$$\psi_A = \frac{1}{2^{1/2}}\{\varphi_n^a\varphi_m^b - \varphi_n^b\varphi_m^a\}. \tag{1.10}$$

Application to the helium atom follows naturally because the perturbation V is now the Coulomb interaction energy between the two atomic electrons. Two series of energy levels emerge, one higher than the other. At this point Heisenberg could say only that rough calculations indicated the ground state to be in the system with higher energy (parahelium) to which corresponds the spatially symmetric wave function ψ_S. The antisymmetric wave function ψ_A

corresponds to orthohelium. Thus far Heisenberg achieved all results without including the electrons' spins.

Further results follow because for arbitrary spin directions of the two electrons each term system (orthohelium or parahelium) is further decomposed depending on the statistical weight of the combined electron spins. There are three possibilities: orthohelium is a spin triplet and parahelium is a spin singlet; orthohelium is a spin singlet and parahelium is a spin triplet; and combinations thereof. In one set the total wave function (space and spin) is antisymmetric, the other symmetric, and combinations of the two. The formalism yields a 'quantum mechanical uncertainty': 'It is an empirical fact that only one system exists and, as far as we can see here, at least qualitatively, it agrees with the helium spectrum; the other system is not realized in nature.' Only the totally antisymmetric wave function is the acceptable solution. Is there something missing from the quantum mechanical formalism which plays the role of selecting one term system from all possible ones?

Here Heisenberg realized the way to include Bose–Einstein statistics and Pauli's exclusion principle: reduction of the statistical weights, or number of possible states, to one is due to Bose–Einstein statistics; that the proper state is antisymmetric is due to Pauli's exclusion principle. Statistics restricts, as well, the concept of electron motion because 'it makes no physical sense to talk about the motion or about the matrix representing the motion of an individual electron' because then one would have to deal with matrix elements for transitions between every possible term system.

In a subsequent paper entitled 'Quantum mechanics' (1926b) Heisenberg came closer to the correct explanation: that reduction of the statistical weights by Bose–Einstein statistics is possible only for indistinguishable particles and 'with electrons, such an equivalence naturally exists'. Consequently no 'intuitive picture' (classical visualization) is possible for the exchange energy. The exchange energy accounts as well for the stability of the helium atom, an insoluble problem in Bohr's atomic physics with its classical concepts of visualization and intuition.

In the 'many-body paper' Heisenberg generalized these results to a system comprising n equal component systems with interactions between two component systems at a time. In general he reasoned thus: For n identical systems there are $n!$ degenerate stationary states, which is just what one would expect from Boltzmann statistics. An interaction lifts the degeneracy leaving $n!$ component systems. Yet nature chooses only one system. Why? Because Bose–Einstein statistics reduces the statistical weight from $n!$ to one, and Pauli's exclusion principle selects the totally antisymmetric wave function. Heisenberg stopped here, having discovered what came to be called Fermi–Dirac Statistics.

Heisenberg recalled that (*AHQP*: 25 February 1963): the 'many-body paper' was 'full of excitement for me because so many new things came up . . . one saw that the mathematical scheme contained new properties, like the symmetry' of the wave function. And as is clear from this paper, continued Heisenberg, 'for quite a considerable time I did mix up Bose statistics with Fermi–Dirac statistics'.

Coincidentally, in the summer of 1926 P. A. M. Dirac completed 'On the theory of quantum mechanics' (1926a) in which totally symmetric wave functions were not excluded. Dirac's overall approach was similar to Heisenberg's but referred to no particular atom.[2] Dirac, too, was somewhat bewildered at the result because quantum mechanics was 'unable to decide which solution is correct [symmetric or antisymmetric]', but the antisymmetric solution agreed with Pauli's exclusion principle. In a footnote added in proof Dirac wrote of Born having informed him that 'Heisenberg has independently obtained results equivalent to these'.

1.6 Transformation theory and word meanings

Despite the success of the new quantum mechanics (for example, calculations of the anomalous Zeeman effect and helium atom spectrum), the physical meaning was unclear of the intermediate manipulations that produced results to be compared with experiment. That is, the mathematical symbols of the quantum mechanics (syntax) did not yet possess unambiguous meanings (semantics).

During the latter part of 1926 into the spring of 1927 at Copenhagen, Bohr and Heisenberg struggled to find a physical interpretation of the quantum mechanics. Heisenberg's review paper of September 1926 'Quantum mechanics' (1926b) enables us to glimpse their struggles, and sets the tone for Heisenberg's own research into quantum electrodynamics. He stressed that our 'ordinary intuition' could not be extrapolated into the atomic realm because the 'electron and the atom possess not any degree of physical reality as the objects of daily experience. Investigation of the type of physical reality which is proper to electrons and atoms is precisely the subject of quantum mechanics'. In Heisenberg's view fundamental problems in quantum mechanics had moved into the realm of philosophy.

After repeated warnings throughout the paper against intuitive interpretations of quantum mechanics, Heisenberg concluded that 'there has been missing in our picture [*Bild*] of the structure of matter any substantial progress toward a contradiction-free intuitive [*anschaulich*] interpretation of experiments'. What could he have meant by a 'contradiction-free intuitive interpretation'? Any reply would have to await Dirac's transformation theory.

Dirac's transformation theory (1926b) provided the mathematical framework missing from Heisenberg's and Pauli's attempts to relate measurements of canonically conjugate variables. Central to Dirac's paper was that Born's probability amplitude is the transformation function between different representations, for example position and energy. Actually Heisenberg had discovered this property of the transformation matrix for the discrete case in his paper 'Fluctuation phenomena and quantum mechanics' (1926c).[3]

Heisenberg's thoughts toward a contradiction-free interpretive framework for quantum mechanics began to crystallize. Throughout he remained focused on the theory's mathematical formalism with its essential discontinuities and nonvisualizability. For both Bohr and Heisenberg linguistic (semantic) difficulties persisted of the same sort as in mid-1925, as Heisenberg described to Pauli in a letter of 23 November 1926 (in Pauli, 1979): 'That the world is continuous I consider more than ever as totally unacceptable. But as soon as it is discontinuous, all our words that we apply to the description of facts are so many c-numbers. What the words "wave" or "corpuscle" mean we know not any more.'

1.7 The uncertainty principle paper

By the end of February 1927 Heisenberg found the connection between measurement of kinematical quantities and Bohr's insistence since 1913 on how unclear the terminology from classical physics becomes when used in a theory for phenomena occurring in a realm beyond sense perceptions. Heisenberg reported these results in his paper (completed March 1927) 'On the intuitive [*anschauliche*] contents of the quantum-theoretical kinematics and mechanics' (1927). The importance of the concept of intuition to Heisenberg is indicated by its inclusion into the title of this classic paper in the history of ideas. By redefining the concept of 'intuition' he found the new 'intuitive interpretation of the various phenomena' for which he had searched in the 1926 paper 'Quantum mechanics'.

Heisenberg's line of argument went as follows: 'The present paper sets up exact definitions of the words position, velocity, energy, etc. (of an electron)'. How can we accomplish this? From our experience with the general theory of relativity we know that the means to extend 'intuitively based' concepts (in the classical meaning of this term) into large space-time regions is 'derivable neither from our laws of thought nor from experiment'. Presently [1927], attempts to obtain an intuitive interpretation of quantum mechanics are full of contradictions because of the 'struggle of opinions concerning discontinuum and continuum theory, particles and waves', which implies that 'it is not possible to interpret quantum mechanics in the customary kinematical terms. [The] necessity of revising kinematical and mechanical concepts appears to follow directly from the basic equations of quantum mechanics; particularly'

$$pq - qp = h/2\pi i, \tag{1.11}$$

from which we 'have good reason to be suspicious about uncritical application of the words "position" and "momentum"'.

Consequently, may we not say that Heisenberg has redefined the concept of intuition [*Anschauung*] with the equations of quantum mechanics? After all, the Kantian notion of intuition contains a concept of visualization that had led physicists astray. Heisenberg separated intuition from visualization by basing all deliberations on unvisualizable particles and essential discontinuities. The mathematics of the quantum mechanics provided the means to extend the concept of intuition into small regions of space-time because it imposed restrictions on perception-laden terms such as position and momentum. What are these restrictions? They are the uncertainty relations which Heisenberg went on to develop with thought experiments that illustrated his view that concepts such as position of an electron and stationary state of an atom derive meaning from experimental measurement.

The well known γ-ray microscope experiment provides a rough derivation of the uncertainty principle for position and momentum

$$p_1 q_1 \sim h, \tag{1.12}$$

where p_1 and q_1 are errors in determination of momentum and position. Heisenberg offered as a substantiation of Eq. (1.12) that '[$p_1 q_1 \sim h$] is the precise expression for the facts which one previously tried to describe by dividing phase space into cells of size h'. In fact, in a letter to Pauli (5 November 1926, in Pauli, 1979), written prior to Dirac's transformation theory, Heisenberg had speculated on a relation such as Eq. (1.12) by analogy with statistical mechanics. In this letter Heisenberg also related to Pauli discussions with Bohr on the possibility that the essential discreteness of quantum mechanics is a clue to the possible discreteness of the space-time metric, which implies the impossibility of measuring jointly or separately momentum and position to any arbitrary degree of accuracy.

Since the uncertainty relations place limits on the accuracy to which initial conditions could be determined then the casual law from classical mechanics that requires both the visualization and continuous development of physical systems is invalid.

From where does the quantum theoretic statistics emerge? Heisenberg preferred an interpretation which he attributed to Dirac: the 'statistic is induced by our experiments'. However, Heisenberg cautioned, we should not conclude that quantum mechanics is 'an essentially statistical theory in the sense that only statistical conclusions can be drawn from specified data'. For example, exact conclusions can be drawn from the conservation laws of energy and momentum. But, owing to the uncertainty relations, speculations that there 'is a "real" world hidden behind the perceived statistical world [are] fruitless and sterile. Physics should describe formally only the connection of perceptions'.

Although Heisenberg's view of physical theory would begin to shift toward rationalism as a result of his research on nuclear theory (see Chapter 5) in 1935, when he seemed to have exhausted all schemes to rid quantum electrodynamics of divergences, he returned to a more positivistic view of physics that emphasized 'only the connection of perceptions' (see the analysis of [10] in Section 4.15).

During the month of February 1927, when Heisenberg wrote the uncertainty principle paper, Bohr was away from Copenhagen on vacation. Upon his return Bohr was critical of Heisenberg's neglect of the wave–particle duality of matter and light which had led Heisenberg to conclude that observational uncertainties were rooted exclusively in the presence of discontinuities. In this way, for example, Heisenberg had reached erroneous conclusions for the γ-ray microscope experiment, among other *Gedanken* experiments. In a 'Note added in proof' Heisenberg (1927) acknowledged comments of this sort, although he made no move to make corrections in the uncertainty principle paper itself.

1.8 Complementarity

On 16 September 1927, at the International Congress of Physics at Como, Italy, Bohr presented his complementarity view, honed in heated discussions with Heisenberg. Since the ordinary intuition could not be extended into the atomic domain, the 'classical mode of description must be generalized' (Bohr, 1928). The usual 'causal space-time description' depended on the smallness of Planck's constant. But in the atomic domain Planck's constant links the measuring apparatus to the system under investigation in a way that 'is completely foreign to the classical theories'. This is how intrinsic statistics enter quantum theory. In the atomic domain the notion of an undisturbed system developing in space and time is an abstraction, and 'there can be no question of causality in the ordinary sense of the word', that is strong causality. Instead of renouncing the causal law like Heisenberg, Bohr linked causality to the predictive powers of the conservation laws of energy and momentum and not to space-time pictures, which he relegated to the role of restricted metaphors.

Bohr went on to reason that, just as the large value of the velocity of light had prevented the realization of the relativity of time, the minuteness of Planck's constant rendered paradoxical the wave–particle duality of matter and light. Since Planck's constant places restrictions on the use of our language in the atomic domain, then so too on our ordinary intuition or visual imagery, which enables us to describe and depict only things that are either continuous or discontinuous but not both. Rather, stressed Bohr, the wave and particle modes of light and matter are neither contradictory nor paradoxical, but complementary in the extreme, that is mutually exclusive. Yet both modes or sides are required for a complete description of the atomic entity. Heisenberg's uncertainty relations are a particular case of complementarity because, for example, the quantities momentum and position are not mutually exclusive. (See Jammer, 1966, Holton, 1973, and Miller, 1988, for further discussions of complementarity.)

These correspondence-limit arguments recurred throughout the development of quantum electrodynamics and received their sharpest formulation in Heisenberg's 1938 paper on the fundamental length [10].

Although Heisenberg agreed with the complementarity principle's restrictions on metaphors from the world of perceptions, he remained wary of them owing to their previous disservices. In a letter of 16 May 1927 to Pauli (in Pauli, 1979), Heisenberg wrote that there are 'presently between Bohr and myself differences of opinion on the word intuitive [*anschauliche*]'. This divergence of opinion widened through Heisenberg's subsequent scientific work.

1.9 Conclusion

Can we not interpret this scenario to be indicative of a switch of Heisenberg's mental representation of knowledge? This switch went beyond merely inverting the Kantian notion of perception in which *Anschauung* is accorded a higher status than *Anschaulichkeit*. Let us summarize the historical results as follows: Until 1924 Bohr and Heisenberg focused on the content of a mental representation – that is what is being represented, which in this case is the *Anschauung* or visualization that classical physics imposed on atomic theory. Starting in 1924, owing to the Bohr–Kramers–Slater version of Bohr's atomic theory, Bohr and Heisenberg began to shift toward emphasis on the format of a representation by permitting the mathematics of the theory to give a purely descriptive representation of the atomic domain. (The format of a mental representation is its encoding.) Yet even after Heisenberg's invention of the new quantum mechanics physicists lamented over loss of visual imagery. In 1927 Heisenberg redefined the concept of intuition by separating it from visualization – that is 'intuition' had no visual content. Rather, visualizability or *Anschaulichkeit* displaced *Anschauung*.

In classical physics visualization and visualizability are the same, but not so in quantum mechanics. Whereas *Anschauung* is a product of our cognitive apparatus, *Anschaulichkeit* pertains to intrinsic attributes of subatomic entities. For example, the electron's spin is an intrinsic property that cannot be perceived and yet which exists regardless of whether or not we set up an experiment to measure it. An early criticism against a 'spinning' electron was that a point on the periphery of the electron would move faster than the velocity of light. In 1927, for Heisenberg and other quantum physicists, visualizability in the atomic domain did not yet possess a depictive or visual component. In contrast, Bohr continued to advocate the usefulness of restricted *Anschauungen*.

In the course of his scientific research in nuclear physics, in 1932 Heisenberg discovered a clue to the depictive mode of visualizability, a mode that would in time enable us to imagine things we have not seen, needless to say within the restrictive framework of our sense perceptions (see Chapter 5, the Epilogue, and Miller, 1984, 1985, 1990, pp. 3–15).

2 Second quantization

2.1 Jordan's 1926 results

Heisenberg wrote in (1929), that 'the existence of the electron' is as unintelligible to the wave mechanical theory as the 'existence of the light quantum' to Maxwell's theory. The fundamental problem for the light quantum was how it could produce interference. The fundamental problem concerning electricity was how quantization of electric charge could be deduced from the Schrödinger wave function because, according to wave mechanics, the total charge on a body is $e\int\Psi^*\Psi\, d^3r$. How could the volume integral of the product of two wave functions be an integer?[4]

Pascual Jordan approached this problem in the *Dreimänner-Arbeit* of 1926, coauthored with Born and Heisenberg, in the section entitled 'Coupled harmonic oscillators. Statistics of wave fields' (Born, Heisenberg and Jordan, 1926 – see letter of Heisenberg to Pauli of 23 October 1926, which confirms that Jordan wrote this section, in Pauli, 1979). Here the line of development begun by Bohr's coupling mechanism is explored further. Jordan cited the investigations of Ehrenfest (1906) and of Debye (1910) noting that neither of these approaches could include the important problem of the 'coupling of distant atoms' because they are semiclassical, mixing classical wave-theoretical notions with light quanta. Consequently, as Einstein (1925) had recognized, although Debye's method leads to Planck's formula, it gives the wrong result for the mean square fluctuations of cavity radiation in a volume element, yielding, instead of the expected two-term result (wave–particle duality of light), only the wave contribution. In (1909) Einstein was able to obtain the correct fluctuation result by using Planck's radiation law and an inversion of Boltzmann's principle.

Jordan proposed to separate the 'theoretical wave-aspect of the problem completely from the theory of light quanta'. The problem was how can interference of wave terms add up to give the correct fluctuation law for cavity radiation?

Jordan's calculation went as follows: He assumed a one-to-one correspondence between cells (in the Bose–Einstein sense) and eigenvibrations. Then he interpreted the quantum numbers of the individual oscillators as the number of light quanta in each cell. He accomplished this by interpreting the quantum number n_k of each oscillator as the number of light quanta with the corresponding frequency ν_k. This weighting of states of the system gives Bose–Einstein statistics. In order to avoid calculational difficulties, Jordan explored the problem of vibrations of a one-dimensional string fixed at both ends.

Consider a string of length l and, for generality, the energy in a part $(0, a)$:

$$E = \frac{1}{2}\int_0^a \left(\left(\frac{\partial u}{\partial t}\right)^2 + \left(\frac{\partial u}{\partial x}\right)^2\right) \mathrm{d}x, \tag{2.1}$$

where appropriate units are used and $u(x, t)$ is the string's transverse displacement. Expressing $u(x, t)$ in a Fourier series as

$$u(x, t) = \sum_{k=1}^{\infty} q_k(t) \sin\left(\frac{k\pi x}{l}\right), \tag{2.2}$$

Eq. (2.1) becomes

$$E = \frac{1}{2}\left[\int_0^a \sum_{j,k=1}^{\infty} \left\{\frac{\mathrm{d}q_j}{\mathrm{d}t}\frac{\mathrm{d}q_k}{\mathrm{d}t} \sin\frac{j\pi x}{l} \sin\frac{k\pi x}{l} + q_j q_k jk\left(\frac{\pi}{l}\right)^2 \cos\frac{j\pi x}{l} \cos\frac{k\pi x}{l}\right\}\right] \mathrm{d}x. \tag{2.3}$$

Calculating the fluctuation in energy from Eq. (2.3) gives

$$\overline{\Delta^2} = \overline{(E - \bar{E}^2)^2} = \frac{\overline{E^2}}{2a}. \tag{2.4}$$

Taking account, however, that in the new quantum mechanics q_k is reinterpreted as a matrix element

$$q_k(n, m) = n_j - m_j = 0 (j \neq k), q_k(n, m) = n_k - m_k = \pm 1.$$

Instead of Eq. (2.4) Jordan obtains the two-term wave–particle duality

$$\overline{\Delta^2} = h\nu\bar{E} + \frac{\overline{E^2}}{Z_\nu V}, \tag{2.5}$$

where Z_ν is the number of characteristic frequencies in the interval $\mathrm{d}\nu$ and V is the volume that contains the string.

In (1925) Ehrenfest pointed out that Einstein's calculations of the wave–particle duality in (1909) for light and in (1925) for atomic entities depended on certain conditions for the additivity of entropies in the volume V of the radiation enclosure where fluctuations are expected and the larger volume of the radiation enclosure that contains V; this additivity is valid only for the classical limit of the Rayleigh–Jeans law. The result in Eq. (2.5) shows that Ehrenfest's criticism does not hold in the new kinematics, at least for this simple case. With confidence Jordan concluded:

> If one bears in mind that the question considered here is actually somewhat remote from the problems whose investigation led to the growth of quantum

> mechanics, the result [Eq. (2.5)] can be regarded as particularly encouraging for the further development of the theory . . .. The basic difference between the theory proposed here and that used hitherto in both instances lies in the characteristic kinematics.

(For further discussion, see Darrigol, 1986.)

2.2 Dirac's quantization of the electromagnetic field

Jordan's method was developed for the electromagnetic field in work Dirac completed by February 1927 whilst at Bohr's institute in Copenhagen. For this purpose Dirac used a more general system than a one-dimensional string.

Consider the Hamiltonian

$$H = H_0 + V, \tag{2.6}$$

where H_0 is the Hamiltonian for an unperturbed system and V is the interaction or perturbation term. The wave function for the total system satisfies the time-dependent Schrödinger equation

$$(H_0 + V)\Psi = i\hbar \frac{\partial \Psi}{\partial t}. \tag{2.7}$$

Taking

$$\Psi = \sum_n a_n \Psi_n, \tag{2.8}$$

where the Ψ_n's are wave functions for the unperturbed system, and the a_n's are expansion coefficients. Then from Eqs. (2.7) and (2.8) as follows:

$$i\hbar \frac{\mathrm{d}a_r}{\mathrm{d}t} = \sum_s V_{rs} a_s, \tag{2.9}$$

$$i\hbar \frac{\mathrm{d}a_r^*}{\mathrm{d}t} = - \sum_s a_s^* V_{rs}, \tag{2.10}$$

where V_{rs} is the matrix element of the perturbation term, and a_r^* is the complex conjugate to a_r. Consequently, a_r and $i\hbar a_r^*$ can be taken to be canonical conjugate variables with a Hamiltonian

$$F_1 = \sum_{r,s} a_r^* V_{rs} a_s, \tag{2.11}$$

that is

$$\frac{\mathrm{d}a_r}{\mathrm{d}t} = - \frac{1}{i\hbar} \frac{\partial F_1}{\partial a_r^*}, \tag{2.12}$$

$$i\hbar \frac{\mathrm{d}a_r^*}{\mathrm{d}t} = \frac{\partial F_1}{\partial a_r}. \tag{2.13}$$

In order to express the a_r's in their real and imaginary parts, Dirac proposed another transformation

$$a_r = N_r^{1/2} \exp(-\mathrm{i}\varphi_r/2), \tag{2.14}$$

$$a_r^* = N_r^{1/2} \exp(\mathrm{i}\varphi_r/2), \tag{2.15}$$

where N_r and φ_r are real and $N_r = |a_r|^2$ is the probable number of systems in the state r. Another transformation sets the equations for the transition quantities a_r and a_r^* into a better form. For this purpose Dirac introduced the quantities b_r and b_r^*, where

$$b_r = a_r \exp(-\mathrm{i}W_r t/\hbar), \tag{2.16}$$

$$b_r^* = a_r^* \exp(\mathrm{i}W_r t/\hbar), \tag{2.17}$$

with W_r as the energy of the unperturbed state Ψ_r, and so

$$\mathrm{i}\hbar \frac{\mathrm{d}b_r}{\mathrm{d}t} = \sum_s H_{rs} b_s, \tag{2.18}$$

where

$$H_{rs} = W_r \delta_{rs} + v_{rs}, \tag{2.19}$$

with

$$V_{rs} = v_{rs} \exp[\mathrm{i}(W_r - W_s)t/\hbar. \tag{2.20}$$

Taking the b_r's as canonically conjugate variables instead of the a_r's means dealing with the Hamiltonian

$$F = \sum_{r,s} b_r^* H_{rs} b_s. \tag{2.21}$$

A contact transformation for the b_r's similar to Eqs. (2.14) and (2.15) yields for F

$$F = \sum_r W_r N_r + \sum_{r,s} v_{rs} N_r^{1/2} N_s^{1/2} \exp[\mathrm{i}(\theta_r - \theta_s)/\hbar], \tag{2.22}$$

with

$$\frac{\mathrm{d}N_r}{\mathrm{d}t} = \frac{\partial F}{\partial t} \text{ and } \frac{\mathrm{d}\theta_r}{\mathrm{d}t} = \frac{\partial F}{\partial N_r}, \tag{2.23}$$

where the first term in Eq. (2.22) is the energy of the assembly and the second term is the energy from the perturbation.

Thus far Dirac had used only ordinary time-dependent perturbation theory. He next assumed that the b_r and b_r^* were q-numbers satisfying the commutation relations

$$[b_r, b_s^\dagger] = \delta_{rs}, \tag{2.24}$$

$$[b_r, b_s] = [b_r^\dagger, b_s^\dagger] = 0 \tag{2.25}$$

(where for q-numbers the notation $b_r^\dagger$ instead of b_r^* is used). Then

$$b_r = (N_r + 1)^{1/2} \exp(-\mathrm{i}\theta_r/\hbar) = \exp(-\mathrm{i}\theta_r/\hbar)\, N_r^{1/2}, \tag{2.26}$$

$$b_s^\dagger = N_r^{1/2} \exp(\mathrm{i}\theta_r/\hbar) = \exp(\mathrm{i}\theta_r/\hbar)(N_r + 1)^{1/2}, \tag{2.27}$$

so that $[N_r, \theta_r] = \hbar/\mathrm{i}$ and the N_r are non-negative integers. Since the procedure in Eqs. (2.24) and (2.25) appears to be one of quantizing the operators b_r again, this process became known as 'second quantization'. The Hamiltonian F becomes

$$F = \sum_{r,s} H_{rs} N_r^{1/2} (N_s + 1 - \delta_{rs})^{1/2} \exp[\mathrm{i}(\theta_r - \theta_s)/\hbar]. \tag{2.28}$$

In the N_r or occupation number representation the Schrödinger equation is

$$\hbar \frac{\partial \Psi(N_1', N_2', \ldots)}{\partial t} = F\Psi(N_1', N_2', \ldots) \tag{2.29}$$

or

$$\mathrm{i}\hbar \frac{\partial \Psi(N_1', N_2', \ldots)}{\partial t} =$$

$$\sum_{r,s} H_{rs} N_r^{1/2} (N_s + 1 - \delta_{rs})^{1/2} \Psi(N_1', N_2', \ldots, N_r' - 1, N_s' + 1, \ldots) \tag{2.30}$$

and so $b_r^\dagger$ (b_r) increases (decreases) by one of the number of systems in the state r.

After demonstrating that F is the Hamiltonian for an interaction with an assembly obeying Bose–Einstein statistics, Dirac went on to consider the interaction between an atom and radiation. The Hamiltonian is

$$H = H_\mathrm{A} + H_\mathrm{R} + V, \tag{2.31}$$

where H_A and H_R are the Hamiltonians for the unperturbed atom and radiation field, respectively, and V is the interaction term, which is

$$V = -\frac{e}{c} \boldsymbol{A} \cdot \dot{\boldsymbol{x}}, \tag{2.32}$$

where $\boldsymbol{A}$ is the vector potential (the scalar potential is taken to be zero) and all calculations are done in the dipole approximation where $\mathrm{e}^{\mathrm{i}\boldsymbol{k}\cdot\boldsymbol{r}} \sim 1$. Thus far the interaction is described from the wave viewpoint in which the vector potential can be written as

$$\boldsymbol{A} \sim b_r \cos\theta_r \boldsymbol{\varepsilon}_r, \tag{2.33}$$

where $\boldsymbol{\varepsilon}_r$ is a unit linear polarization vector, and so

$$F = H_\mathrm{A} + \sum_r \hbar\omega N_r + \frac{e}{c} \sum_r \mathbf{x}_r \cdot \boldsymbol{\varepsilon}_r N_r^{1/2} \cos\theta_r. \tag{2.34}$$

Making N_r and θ_r q-numbers operating on the wave function $\Psi(N_1, N_2, \ldots)$ involves the replacement

$$N_r^{1/2} \cos\theta_r \to \frac{1}{2}\{N_r^{1/2} \exp(\mathrm{i}\theta_r/\hbar) + (N_r + 1)^{1/2} \exp(-\mathrm{i}\theta_r/\hbar)\}, \tag{2.35}$$

which completes the transformation from waves to particles: 'There is thus a complete harmony between the wave and light quantum descriptions of the interaction'. Dirac completed this paper in Copenhagen and it was communicated by Bohr to the Royal Society (received 2 February 1927). Here we may inquire whether the 'complete harmony between waves and particles' offered by Dirac's formalism influenced Bohr's thoughts toward complementarity.

Dirac went on to formulate a procedure to calculate the absorption and emission of radiation, permitting him to derive the Einstein A and B coefficients as well. Toward future considerations in quantum field theory another noteworthy point about Dirac's paper is his description of photon absorption and emission. When a photon is absorbed it is in its zeroth stationary state and so its momentum and energy are zero. When a photon is emitted it jumps from the zero state 'so that it appears to have been created'. Since there are an infinite number of photons in the zeroth state, one expects an infinite occupation number N_0. Although Dirac never further developed this representation for the zero photon state, might it have influenced his concept of the vacuum as a sea of negative energy electrons?

2.3 Jordan's quantization of bosons and fermions

Jordan extended Dirac's quantization methods to Bose particles with mass in October 1927 (Jordan and Klein, 1927), and then to Fermi–Dirac particles in January 1928 (Jordan and Wigner, 1928). For agreement with Pauli's exclusion principle the Fermi–Dirac case required invention of anticommutators.

Of particular interest for further considerations is the Jordan–Klein method for dealing with the Coulomb self-interaction energy of an electron with charge distribution $\rho(r, t)$

$$E_{\mathrm{Coul}} = \frac{1}{2}\int\int \frac{\rho(r, t)\rho(r', t)}{|r - r'|}\, \mathrm{d}^3 r\, \mathrm{d}^3 r', \tag{2.36}$$

which becomes in quantum mechanics

$$E_{\mathrm{Coul}} = \frac{1}{2}\int\int \frac{\Psi^\dagger(r, t)\Psi(r, t)\Psi^\dagger(r', t)\Psi(r', t)}{|r - r'|}\, \mathrm{d}^3 r\, \mathrm{d}^3 r', \tag{2.37}$$

where Jordan and Klein took the Ψ's to be operators in occupation number space*. They noted that, although Eq. (2.37) is the form that 'naturally occurs . . . this is not correct' because all creation operators must stand to the left of all annihilation operators. Since they dealt in this paper with bosons

$$\Psi(r, t)\Psi^{\dagger}(r', t) = \delta(r - r') + \Psi^{\dagger}(r', t)\Psi(r, t) \tag{2.38}$$

and the corrected energy is

$$E_{\text{corr}} = E_{\text{Coul}} - \frac{e^2}{2}\int\int \frac{\Psi^{\dagger}(r', t)\Psi(r, t)}{|r - r'|}\delta(r - r')\,\mathrm{d}^3 r\,\mathrm{d}^3 r', \tag{2.39}$$

where the last term is the infinite self-energy of a point charge. E_{corr} turns out to be the same for bosons and fermions (for example, see Schiff, 1968).

Jordan and Klein (1927) wrote that the corrected expression in Eq. (2.39) has an 'intuitive [*anschauliche*] meaning' owing to its 'suitable analogy to classical theory' in which the interaction energy of a system of mass points is

$$\frac{e^2}{2}\sum_{k\neq l} G(r_k, r_l) = \frac{e^2}{2}\sum_{k,l} G(r_k, r_l) - \frac{e^2}{2}\sum_{k} G(r_k, r_l), \tag{2.40}$$

where $G(r_k, r_l) = |r_k - r_l|^{-1}$. Suppose that $N(r')\mathrm{d}^3 r' = \Psi^{\dagger}(r', t)\Psi(r', t)\mathrm{d}^3 r'$ of these mass points are situated within the volume element $\mathrm{d}^3 r'$, then the sums in Eq. (2.40) can be replaced by the integrals

$$\frac{e^2}{2}\int\int G(r', r'')N(r')N(r'')\mathrm{d}^3 r'\mathrm{d}^3 r''$$
$$- \frac{e^2}{2}\int G(r, r)N(r)d^3 r. \tag{2.41}$$

Jordan and Klein emphasized that the noncommutative property of the quantum mechanical wave functions 'remarkably' enable replacement of the difference between double and single volume integrals in Eq. (2.41) by a single double volume integral. Consequently quantum mechanics (Eq. (2.39)) expresses in 'analytically simple form that the "proper field [*Eigenfeld*]" of an electron does not act upon it in the same fashion as the "external field" – a difference which appeared to be expressable exactly in classical theory only very unsatisfactorily and only with difficulty'.

The technique in Eq. (2.39) is a precursor to the subtraction methods introduced into quantum electrodynamics by Dirac in 1933 [3] (discussed in Section 4.7). We will see that in retrospect Heisenberg was doubly correct

* As a historical comment on notation, regarding the use of superscript daggers for the adjoint of a field operator, I quote from Jordan and Klein: 'The quantity $b_s^{\dagger}$ [where $b_s^{\dagger}$ is the quantity in Eq. (2.27)] emerges from b_s not only by replacing i by −i but also by inverting the sequence of multiplications; for this reason we should use the designation $b^{\dagger}$ corresponding to a recently made proposal to characterize such "adjoint" quantities by a dagger.'

when he perceived that reordering of creation and annihilation operators, so that creation operators stand to the left of annihilation operators (which he referred to as the 'Klein–Jordan trick'), could provide the path to removal of the electron's divergent self-energy. But the full formulation of this technique would also require Heisenberg's invention of the S-matrix in (1943a) followed by Gian Carlo Wick's (1950) method of the normal product expansion of the S-matrix into terms each one equivalent to a Feynman diagram (see Dyson, 1951a,b; in less elegant form, without the Wick expansion, this result is present also in Dyson, 1949a,b).

2.4 Jordan and Pauli's relativistic quantization of charge-free electromagnetic fields

In Dirac's quantization method for the electromagnetic field, and in Jordan's extensions to particles with mass, wrote Jordan and Pauli, the 'time coordinate is always specially distinguished from the space coordinates, and the results are not relativistically invariant' (Jordan and Pauli, 1928).

Before formulating a relativistically invariant quantization method for the electromagnetic field, Jordan and Pauli turned to a nagging problem of their quantization procedure, namely the zero-point energy of the harmonic oscillator. The harmonic oscillator Hamiltonian

$$H = \frac{p^2}{2m} + \frac{m\omega^2 q^2}{2} \tag{2.42}$$

has eigenvalues

$$E_n = (n + \tfrac{1}{2})h\nu. \tag{2.43}$$

Consequently the total energy density is infinite even for the $n = 0$ state. But this energy is spurious because, contrary to the case of the proper vibrations in a crystal lattice, the zero-point energy for radiation is nonphysical because it can be neither 'absorbed nor scattered nor reflected'. So in a 'more satisfactory interpretation' it should not exist at all.[5]

Jordan and Pauli went on to introduce a canonical transformation that eliminates the zero-point energy for a single oscillator, expressing the hope that 'along this route' the zero-point energy could be eliminated for the case of an infinite number of oscillators.[6]

At this point it is convenient to use notation that made its appearance in the 1930s. In Gaussian unrationalized units the Maxwell–Lorentz equations for a system of radiation and charges are

$$\operatorname{curl} \boldsymbol{E} = -\frac{1}{c}\frac{\partial \boldsymbol{B}}{\partial t}, \tag{2.44}$$

$$\operatorname{curl} \boldsymbol{B} = \frac{1}{c}\frac{\partial \boldsymbol{E}}{\partial t} + \frac{4\pi\rho \boldsymbol{v}}{c}, \tag{2.45}$$

$$\operatorname{div} \boldsymbol{E} = 4\pi\rho, \tag{2.46}$$

$$\operatorname{div} \boldsymbol{B} = 0. \tag{2.47}$$

The quantized vector potential is

$$\boldsymbol{A}(\boldsymbol{x}, t) = \frac{1}{V^{1/2}} \sum_{k,\alpha} \left(\frac{2\pi\hbar c^2}{\omega_k} \right)^{1/2} [a_{k,\alpha} \boldsymbol{\varepsilon}_\alpha \mathrm{e}^{ik\cdot x} + a^\dagger_{k,\alpha} \boldsymbol{\varepsilon}_\alpha \mathrm{e}^{-ik\cdot x}], \tag{2.48}$$

where $V = L^3$ is the cubical quantization volume, the k's satisfy the periodic boundary condition at the walls (that is, k_x, k_y, $k_z = 2\pi n/L$, where $n = \pm 1$, $\ldots$, $\boldsymbol{\varepsilon}_\alpha$ is a unit linear polarization vector), $k \cdot x = \boldsymbol{k} \cdot \boldsymbol{x} - kct$, where $k = \omega_k/c$, as a limit

$$\frac{1}{V^{1/2}} \sum_k \rightarrow (2\pi)^{-3/2} \int \mathrm{d}^3 k. \tag{2.49}$$

The $a_{k,\alpha}$ and $a^\dagger_{k,\alpha}$ are annihilation and creation operators for photons of momentum k and polarization α, respectively, that satisfy commutation relations

$$[a_{k,\alpha}, a^\dagger_{k',\alpha}] = \delta_{kk'}\delta_{\alpha\alpha}, \tag{2.50}$$

$$[a_{k,\alpha}, a_{k',\alpha'}] = [a^\dagger_{k,\alpha}, a^\dagger_{k',\alpha'}] = 0. \tag{2.51}$$

The creation and annihilation operators transform a state $|n_{k_1,\alpha_1}, \ldots, n_{k_i,\alpha_i}, \ldots\rangle$ in occupation number space as follows:

$$a^\dagger_{k_i,\alpha_i} |n_{k_1,\alpha_1}, \ldots, n_{k_i,\alpha_i}, \ldots\rangle = (n_{k_i,\alpha_i} + 1)^{1/2} |n_{k_1,\alpha_1}, \ldots, n_{k_i,\alpha_i} + 1, \ldots\rangle, \tag{2.52}$$

$$a_{k_i,\alpha_i} |n_{k_1,\alpha_1}, \ldots, n_{k_i,\alpha_i}, \ldots\rangle = (n_{k_i,\alpha_i} - 1)^{1/2} |n_{k_1,\alpha_1}, \ldots, n_{k_i,\alpha_i} - 1, \ldots\rangle. \tag{2.53}$$

There is a number operator

$$N_{k,\alpha} = a^\dagger_{k,\alpha} a_{k,\alpha}, \tag{2.54}$$

with eigenvalues 0, 1, 2, $\ldots$, that designate the number of photons of type k, α, that is

$$N_{k_i,\alpha_i} |n_{k_1,\alpha_1}, \ldots, n_{k_i,\alpha_i}, \ldots\rangle = n_{k_i,\alpha_i} |n_{k_1,\alpha_1}, \ldots, n_{k_i,\alpha_i}, \ldots\rangle. \tag{2.55}$$

The Hamiltonian in the wave theory of the electromagnetic field is

$$H = \frac{1}{8\pi} \int (\boldsymbol{E}^2 + \boldsymbol{B}^2)\, \mathrm{d}^3 r, \tag{2.56}$$

which becomes in second quantization

$$H = \sum_{k,\alpha}\left(N_{k,\alpha} + \frac{1}{2}\right)\hbar\omega_k. \tag{2.57}$$

Except for the zero-point energy in Eq. (2.57) and the polarizations, the results of this sketch of second quantization hold for bosons, and, with commutators replaced by anticommutators, for electrons too.

Jordan and Pauli moved to 'define a new relativistically invariant δ-function' in order to set up relativistically invariant commutators for $\boldsymbol{E}$ and $\boldsymbol{B}$. Their quantization procedure was more general than those of Dirac (1927a) and Jordan and Klein (1927) because Pauli and Jordan quantized the field strengths $\boldsymbol{E}$ and $\boldsymbol{B}$ themselves.

Expressing $\boldsymbol{E}$ and $\boldsymbol{B}$ in terms of $\boldsymbol{A}$ they obtained

$$[E_j(\boldsymbol{r}, t), E_k(\boldsymbol{r}', t')] = 4\pi i\hbar c\left[\frac{\delta_{jk}}{c^2}\frac{\partial}{\partial t}\frac{\partial}{\partial t'} - \frac{\partial}{\partial r_j}\frac{\partial}{\partial r_k}\right] \cdot D(\boldsymbol{r} - \boldsymbol{r}', t - t'), \tag{2.58}$$

$$[E_j(\boldsymbol{r}, t), E_k(\boldsymbol{r}', t')] = [B_j(\boldsymbol{r}, t), B_k(\boldsymbol{r}', t')], \tag{2.59}$$

$$[E_x(\boldsymbol{r}, t), B_y(\boldsymbol{r}', t')] = -[B_x(\boldsymbol{r}, t), E_y(\boldsymbol{r}', t')] = -4\pi i\hbar\frac{\partial}{\partial z}\frac{\partial}{\partial t'}D(\boldsymbol{r} - \boldsymbol{r}', t - t'), \tag{2.60}$$

with cyclic permutations of x, y and z in Eq. (2.60). The Lorentz invariant Jordan–Pauli D-function is

$$D(\boldsymbol{r}, t) = \frac{1}{4\pi r}\{\delta(r + ct) - \delta(r - ct)\}, \tag{2.61}$$

with the following properties:

$$\Box D = 0, \tag{2.62}$$

where the D'Alembertian operator

$$\Box = \Delta - \frac{1\partial^2}{c^2\partial t^2}, \tag{2.63}$$

with Δ the Laplacian, and

$$D(\boldsymbol{r}, 0) = 0, \tag{2.64}$$

$$D(\boldsymbol{r}, t) = 0 \text{ for } t < r/c, \tag{2.65}$$

$$\left.\frac{\partial D(\boldsymbol{r}, t)}{\partial t}\right|_{t=0} = c\delta(\boldsymbol{r}), \tag{2.66}$$

that is the D-function vanishes outside of the light cone and has singularities on the light cone. Consequently electric and magnetic fields can be measured to arbitrary accuracy at two points with a space-like separation. These results show also that the quantized electromagnetic field propagates at the velocity of light.

Jordan and Pauli rightly expected that their work would be the limiting case of a more general framework that includes charged particles.

3 Photons and relativistic electrons

3.1 The Dirac equation

Until Dirac's electron theory, the only relativistically invariant wave equation was that of Klein and Gordon (Gordon, 1926, and Klein, 1926).[7]

$$\left(\Box - \frac{m^2c^2}{\hbar^2}\right)\psi = 0, \tag{3.1}$$

with a particle density

$$\rho = \psi^* \frac{\hbar}{\mathrm{i}} \frac{\partial \psi}{\partial t} - \frac{\hbar}{\mathrm{i}} \frac{\partial \psi^*}{\partial t} \psi, \tag{3.2}$$

and a Hamiltonian density

$$H = \hbar^2 \left|\frac{\partial \psi}{\partial t}\right|^2 + \hbar^2 c^2 |\nabla \psi|^2 + m^2 c^4 |\psi|^2. \tag{3.3}$$

Contrary to the standard introductions in physics textbooks to the Dirac equation, Dirac's initial motivations had little to do with the nonpositive definiteness of the particle density in Eq. (3.2), which could not be interpreted as a probability density (see Kragh, 1981, and Moyer, 1981). Nor did the possibility of negative energy states deter Dirac because these states exist in any relativistic theory, owing to the relativistic equation for energy

$$E = \pm(p^2c^2 + m^2c^4)^{1/2}. \tag{3.4}$$

Whereas classical mechanics excludes negative energies as unphysical, in quantum mechanics transitions can occur between $+mc^2$ and $-mc^2$.

What most likely did motivate Dirac was the need for a relativistic equation with a first-order time derivative in order to satisfy the quantum mechanical transformation theory. Yet, owing to the negative energy states, Dirac, himself, at first considered his theory to be provisional (1928). Nevertheless, the theory had enormous initial successes (natural emergence of electron spin and correct Sommerfeld fine structure formula).

The Dirac relativistic equation for an electron of mass m and charge $-|e|$ is

$$\left(\not{\partial} + \frac{mc}{\hbar}\right)\psi = 0, \tag{3.5}$$

where ψ is a four-component spinor with two components corresponding to negative energy states, $\not{\partial} = \gamma_\mu \partial_\mu$, with γ_μ the 4×4 Dirac γ-matrices in the representation

$$\gamma_k = \begin{pmatrix} 0 & -\mathrm{i}\sigma_k \\ \mathrm{i}\sigma_k & 0 \end{pmatrix} \quad \text{and} \quad \gamma_4 = \begin{pmatrix} I & 0 \\ 0 & -I \end{pmatrix}. \tag{3.6}$$

The σ_k are 2×2 Pauli spin matrices and I is the 2×2 identity matrix; the γ-matrices anticommute

$$\{\gamma_\mu, \gamma_\nu\} = 2\delta v_{\mu\nu}, \tag{3.7}$$

the particle density is

$$\rho = \psi^* \psi, \tag{3.8}$$

and the Hamiltonian density for a free electron is

$$H = \psi^*[-\mathrm{i}\hbar c \boldsymbol{\alpha} \cdot \boldsymbol{\nabla} + mc^2 \beta]\psi, \tag{3.9}$$

where

$$\alpha_k = \begin{pmatrix} 0 & \sigma_k \\ \sigma_k & 0 \end{pmatrix} \quad \text{and} \quad \beta = \begin{pmatrix} I & 0 \\ 0 & -I \end{pmatrix}. \tag{3.10}$$

Electromagnetic potentials are inserted by the replacement

$$\partial_\mu \to \partial_\mu + \frac{\mathrm{i}e}{\hbar c} A_\mu, \tag{3.11}$$

in which $A_\mu = (\boldsymbol{A}, \mathrm{i}\varphi)$, where $\boldsymbol{A}$ is the vector potential and φ is the scalar potential for a particle of charge $-|e|$.

Until the advent of covariant perturbation theory in 1946, ordinary time-dependent perturbation theory was usually used to treat self-energy and collision problems. For this purpose the Dirac equation in Eq. (3.5) was rewritten to look like the Schrödinger equation:

$$H\psi = \mathrm{i}\hbar \frac{\partial}{\partial t} \psi. \tag{3.12}$$

So, from Eq. (3.9) the Dirac equation for an electron in an electromagnetic field becomes

$$\left\{ \boldsymbol{\alpha} \cdot \left(\frac{\hbar}{\mathrm{i}} \boldsymbol{\nabla} + e\boldsymbol{A} \right) + mc^2 \beta + e\varphi \right\} \psi = \mathrm{i}\hbar \frac{\partial}{\partial t} \psi \tag{3.13}$$

(other variations of perturbation theory were developed by Heisenberg and Pauli in (1929 and 1930) and by Heisenberg in (1931 and 1934a), which we discuss in a moment).

Owing to the negative energy states, Heisenberg deemed the Dirac equation to be the 'saddest chapter in modern physics' (letter to Pauli, 31 July 1928, in Pauli, 1985). Although the negative energies were considered to be extraneous, they were ubiquitous. For example, wrote Heisenberg, in atoms the 'spontaneous radiation transitions $+mc^2$ to $-mc^2$ are more abundant than any other spontaneous transition'. And in Dirac's dispersion theory, the classical Thomson limit for light scattered by an electron emerges from the negative energy states only (that is in the limit $\hbar\omega \ll mc^2$, where ω is the frequency of the incident light).[8]

But Heisenberg and Pauli took Dirac's work seriously enough to begin formulating a quantum electrodynamics, that is a relativistic quantum theory for the interaction between radiation and electrons.

Among the new results that appeared before publication of Part I of the Heisenberg–Pauli opus (1929) are the following.

By October 1928 Klein and Y. Nishina used Dirac's equation and semiclassical radiation theory (unquantized radiation field), along with Dirac's time-dependent second-order perturbation theory to calculate the Compton effect (including the 'physically meaningless negative energy states'; Klein and Nishina, 1928).

By the end of December 1928 Klein (1928) completed his work on the peculiarities of electrons that are reflected at a potential step. Owing to relativistic kinematics we have the following situation for the potential step $V(x) = 0$ for $x < 0$ and $V(x) = +P$ for $x > 0$: If $P > E + m_0c^2$ (where E is the energy of an electron approaching from $x < 0$) an electron can penetrate the barrier and emerge into $x > 0$ with a negative kinetic energy and with velocity directed oppositely to its momentum. Klein concluded ominously that the energy 'difficulty . . . can occur with purely mechanical problems where there is no question of radiative processes'. Jokingly, Pauli referred to Klein's results as a 'scandal' (letter of Pauli to Klein, 18 February 1929, in Pauli, 1985).[9]

3.2 Heisenberg and Pauli on quantum electrodynamics, 1929

Heisenberg and Pauli (1929) set up a theory of the interaction of light and matter 'by considerations based on correspondence limits . . . . The correspondence-like analog to the theory attempted here will be on the one hand Maxwell's theory and on the other hand the wave equation of the single-electron problem reinterpreted in the sense of the classical continuum theory'.

They accomplished this program by inventing a method for field quantization to which they were guided by classical Lagrangian field theory.

(1) Write the system's Lagrangian density L as

$$L = L(\psi_\alpha, \text{grad}\,\psi_\alpha, \dot{\psi}_\alpha), \tag{3.14}$$

where α labels the components of the field.

(2) Calculate the canonical momentum π_α to each field component

$$\pi_\alpha = \frac{\partial L}{\partial \dot{\psi}_\alpha}. \tag{3.15}$$

(3) Write the system's Hamiltonian density H

$$H(\pi_\alpha, \psi_\alpha) = \sum_\alpha \pi_\alpha \dot{\psi}_\alpha - L. \tag{3.16}$$

(4) Require that π_α and ψ_α are q-numbers, that is

$$[\pi_\alpha, \psi_\alpha] = \frac{\hbar}{\mathrm{i}}. \tag{3.17}$$

where the commutator is evaluated at equal times for different spatial points.

To the problem of whether one uses commutators or anticommutators to quantize q-numbers, they replied:

> It is apparent that from the standpoint of wave quantization and of the relativistically invariant treatment of the many-body problem, the two types of solutions, namely Bose–Einstein statistics on the one hand, and the exclusion principle on the other hand, will also always appear as formally equivalent. Thus, no satisfactory explanation can be given why nature prefers the second possibility.

Heisenberg and Pauli regarded the step of quantizing with either commutators or anticommutators to be a 'well known peculiarity',[10] and proceeded to do all calculations with commutators and anticommutators. As Pauli (1985) wrote to Klein on 18 February 1929 concerning either method of quantization, 'we obtain not any information at all on the inner foundation of my exclusion principle'. This quest for an 'inner foundation' was central to Pauli's thoughts during the 1930s.

Two difficulties emerged for the first time from the Heisenberg–Pauli paper: The first difficulty was that quantization of Maxwell's equations turned out not to be as straightforward as they at first thought. The Lagrangian density for the electromagnetic field without sources is

$$L = \frac{1}{8\pi}(\boldsymbol{B}^2 - \boldsymbol{E}^2), \tag{3.18}$$

or in terms of potentials

$$L = \frac{1}{8\pi}\left[(\mathrm{curl}\,\boldsymbol{A})^2 - \left(\frac{1}{c}\frac{\partial \boldsymbol{A}}{\partial t} - \nabla\varphi\right)^2\right]. \tag{3.19}$$

Since $\mathrm{d}A_4/\mathrm{d}t$ $(= \mathrm{i}\,\mathrm{d}\varphi/\mathrm{d}t)$ does not appear in L, then its canonical momentum $\pi_4 = 0$ and the commutator

$$[\pi_4, A_4] = 0,$$

instead of $(\hbar/\mathrm{i})$.[11]

Consequently the quantization procedure failed. This was a tremendous setback to the program. As had been the case in 1924, when he failed to solve the Zeeman effect, and in the spring of 1925, when the old Bohr atomic physics completely broke down with no alternative in sight, Pauli once again despaired. In the 1924 crisis he delved into the Hamburg underground café culture; in 1925 he considered taking up movie directing; and in 1929 he began to write a Utopian novel (letter to Peierls of 18 June 1929, in Pauli, 1985). Heisenberg, however, thrived in periods of flux and resolved this difficulty.

Heisenberg's '*Kunstgriff*', as Pauli referred to it to Peierls, was to add to the Lagrangian density the term

$$L' = \frac{\varepsilon}{2}\left(\frac{\partial \varphi_\alpha}{\partial x_\alpha}\right)^2, \tag{3.20}$$

which preserves gauge and relativistic invariance and also gauge invariance with the replacement $\varphi'_\alpha = \varphi_\alpha + \partial\lambda/\partial x_\alpha$, imposing on the function λ the auxiliary condition $\Box\lambda = 0$. (Repeated Greek indices indicate summation from 1 to 4 and repeated Latin indices mean summation from 1 to 3.) The quantity ε is set equal to zero at the end of any calculation.

In Part II of their paper, Heisenberg and Pauli (1930) removed the 'aesthetic defect' of the ε term by demonstrating that in a relativistically invariant manner one can always choose a gauge so that $\varphi = 0$. Consequently all calculations can be done in the Coulomb gauge without supplementary terms.

The second difficulty was that there arises an infinite self-energy term for the electron, which is the 'interaction of the particle with itself'. Heisenberg and Pauli explored this term further by developing a perturbation method alternative to time-dependent perturbation theory. They wrote the Schrödinger equation in occupation number space by expressing the q-number electron wave function and the electromagnetic field in creation and annihilation operators. Schrödinger's equation thus became

$$(-E + H_{\text{atom}} + H_{\text{rad}})\Phi = -H_{\text{int}}\Phi, \tag{3.21}$$

where $\Phi(N_1, N_2, \ldots; M_1, M_2, \ldots)$ is the state comprising N_s electrons in states s, M_r photons in states r and

$$H_{\text{int}} = -\mathrm{i}e\bar{\psi}\gamma_\mu\psi A^\mu, \tag{3.22}$$

where $\bar{\psi} = \psi^\dagger\gamma_4$. Expressing Φ as $\Phi = \Phi_0 + \Phi_1 + \cdots$, where Φ_0 is the wave function of the unperturbed system of electrons, that is

$$\Phi_0 = \delta_{N_1 N_1^0}\delta_{N_2 N_2^0}\ldots\delta_{M_1 0}\delta_{M_2 0}\ldots.$$

where N_r^0 is the number of electrons in the unperturbed state r, and there are no photons in the unperturbed system of electrons. Then an iteration procedure can commence with the additional substitution of $E = E_0 + E_1 + E_2 + \cdots$. The term E_1 vanishes because it is proportional to the diagonal matrix element of H_{int}. E_2 contains the electron's Coulomb self-energy. The 'Klein–Jordan trick' however, does not work for the electron in the presence of an electromagnetic field because the A_k commute with the electron wave function ψ.[12]

In Part I Heisenberg and Pauli (1929) simply discard the infinite Coulomb self-energy as an irrelevant additive infinite constant, just like the zero-point energy of the vacuum; neither of these effects, they wrote, 'correspond to reality'. They omitted any physical discussion of the negative energy states, 'since these transitions [that is, transitions from $+mc^2$ to $-mc^2$] undoubtedly do not occur in reality'.[13] The negative energy states were an 'inconsistency of the theory being presented here, which must be accepted as long as the Dirac difficulty is unexplained'.

In the improved formulation of their quantum electrodynamics in Part II, Heisenberg and Pauli were less sure about neglecting the infinite Coulomb self-energy, the presence of which 'in many cases will make application of the theory impossible'.

Despite these problems, they believed that any future theory of quantum electrodynamics would have 'essential traits in common with the theory attempted here'.[14]

It was Pauli's belief that the difficulties and shortcomings of the Heisenberg–Pauli formulation of quantum electrodynamics indicated that the 'natural goal of the force and range of correspondence thinking on the basis of wave mechanics had been reached' (letter to Peierls of 18 June 1929).

Further difficulties ensued. In a sequel paper, J. Robert Oppenheimer, who had worked with Heisenberg and Pauli on Part II, found a new contribution to the self-energy of the electron. By rigorously carrying out Heisenberg and Pauli's iteration procedure from Part II of their paper, Oppenheimer (1930b) identified another self-energy for the bound electron, which led to the false prediction of infinite displacements of energy levels and which could not be disregarded merely as an irrelevant infinite additive constant.[15] Like Heisenberg and Pauli, Oppenheimer discarded the infinite Coulomb self-energy.

Oppenheimer's calculation was redone more directly and clearly by Ivar Waller (1930b), who demonstrated that, for free electrons moving with momentum p, second-order perturbation theory gives an additional divergence to the electron's self-energy that has a leading term

$$W(p) \sim \hbar \int_0^\infty k \, \mathrm{d}k \tag{3.23}$$

and so diverges quadratically, where k is the energy of an intermediate state photon. The quantity $\hbar$ is exhibited explicitly to emphasize that the self-energy in Eq. (3.23) is a quantum effect. It arises in second-order perturbation theory from the electron's own vector potential (here expressed in the Coulomb or radiation gauge). This so-called transverse self-energy is in addition to the Coulomb self-energy that is expected from classical electromagnetic theory and which diverges linearly like $1/r_0$, where $r_0 = e^2/mc^2$ is the classical electron's radius. Oppenheimer and Waller pointed out that the divergent self-energy was not directly connected with the negative energy states.

3.3 The electron's mass in classical and quantum electrodynamics

Owing to reliance by Heisenberg and Pauli on correspondence-limit procedures in quantum electrodynamics, it is useful to pause here and compare how the mass of the electron is calculated in classical and quantum electrodynamics.

Lack of success in deducing the equations of electromagnetism and also the properties of light from Newtonian mechanics – a mechanical world-picture – led Wilhelm Wien (1900) to propose research toward an electromagnetic world-picture. In this program the Maxwell–Lorentz equations are taken as axiomatic and the electron's mass is to be deduced from its own (or self-) electromagnetic field. Consequently, like a self-induction effect, the moving-electron's mass should be a function of its velocity (relative to the laboratory) that increases without limit as the electron's velocity approaches the velocity of light. Empirical data available in 1902 gave just this dependence of mass on velocity.

The subsequent theories of Max Abraham (1902, 1903), who set the mathematical basis for this sort of approach, and H. A. Lorentz (1904) were formulated as follows: The goal of the electromagnetic world-picture is to deduce the electron's mass from the Lagrangian L^{e} containing only the electron's self-electromagnetic fields, where

$$L^{\mathrm{e}} = \frac{1}{8\pi}\int (\boldsymbol{B}^2 - \boldsymbol{E}^2)\,\mathrm{d}V, \tag{3.24}$$

$\boldsymbol{E}$ and $\boldsymbol{B}$ are the electron's self-electric and self-magnetic fields and the integral is carried out over the volume of the electron. The electron's momentum is deduced from Eq. (3.24) by the usual Lagrangian methods ($G^{\mathrm{e}} = \mathrm{d}L/\mathrm{d}v$) with certain assumptions on the electron's constitution and charge distribution.

To complete the deduction of the electron's mass requires recourse to Newton's second law

$$\boldsymbol{F}^{\mathrm{e}} + \boldsymbol{F}_{\mathrm{ext}} = m_0\boldsymbol{a}, \tag{3.25}$$

where the resultant force on the electron is the sum of the external forces $\boldsymbol{F}_{\text{ext}}$ and the self-electromagnetic force $\boldsymbol{F}^{\text{e}}$, which is the Lorentz force

$$\boldsymbol{F}^{\text{e}} = \int \rho \left(\boldsymbol{E} + \frac{\boldsymbol{\nabla}}{c} \times \boldsymbol{B} \right) \mathrm{d}V, \tag{3.26}$$

where ρ is the electron's charge density, and m_0 is the electron's mechanical (inertial) mass, which is assumed to be zero in the electromagnetic world-picture. Consequently Eq. (3.25) becomes

$$\boldsymbol{F}_{\text{ext}} = -\boldsymbol{F}^{\text{e}}. \tag{3.27}$$

Expansion of the self-fields in Eq. (3.26) expressed as retarded potentials yields

$$\boldsymbol{F}^{\text{e}} = -\alpha_1 \boldsymbol{a} + \alpha_2\, \mathrm{d}\boldsymbol{a}/\mathrm{d}t + \text{(terms proportional to higher-order time derivatives of the acceleration } \boldsymbol{a}\text{)}, \tag{3.28}$$

where the term α_1 is dependent on the electron's structure, whereas α_2 is not. The term α_2 is the radiation-reaction force, that is the force exerted on the electron by its self-fields. Then, from Eq. (3.27),

$$\boldsymbol{F}_{\text{ext}} = \alpha_1 \boldsymbol{a} - \alpha_2\, \mathrm{d}\boldsymbol{a}/\mathrm{d}t + \text{(terms proportional to higher-order time derivatives of the acceleration } \boldsymbol{a}\text{)}. \tag{3.29}$$

Maintaining an equivalent of Newton's second law in the electromagnetic world-picture requires substantiation that $\alpha_2\, \mathrm{d}\boldsymbol{a}/\mathrm{d}t \ll \alpha_1 \boldsymbol{a}$. This proof was first accomplished by Abraham (1903). A dimensional analysis by Poincaré in (1906) will suffice. The quantity α_1 is proportional to the electron's electromagnetic (i.e. electrostatic) rest mass, which is

$$\mu_0^{\text{e}} = \frac{e^2}{r_0 c^2}, \tag{3.30}$$

where r_0 is the electron's radius (of the order of 10^{-15} m), and certain numerical factors have been omitted from Eq. (3.30) which are essential to a complete historical analysis of the electron's electromagnetic mass (see Miller, 1981). The quantity α_2 can be written as

$$\alpha_2 = e^2/c^3. \tag{3.31}$$

Then

$$\alpha_1 a \sim \frac{e^2}{r_0 c^2}\frac{d}{T^2}, \qquad \alpha_2 \frac{\mathrm{d}a}{\mathrm{d}t} \sim \frac{e^2}{c^3}\frac{d}{T^3}, \tag{3.32}$$

where d is the distance over which the electron accelerates from velocity v_1 to velocity v_2 during a time interval T. In order for $\alpha_2\, \mathrm{d}a/\mathrm{d}t \ll \alpha_1 a$, from Eq. (3.32) it follows that

$$\frac{r_0}{cT} \ll 1. \tag{3.33}$$

Consequently the condition for truncating the series in Eq. (3.29) at the first term is that the time interval T over which the electron is accelerated from velocity v_1 to velocity v_2 exceeds the time for light to traverse the electron's radius. This is the approximation of quasi-stationary motion, which permitted Abraham and Lorentz to calculate the electromagnetic mass by assuming that the electron's self-fields were those for a charge moving with a uniform velocity that happens to be the electron's instantaneous velocity. Consequently the time rate of change of the momentum calculated from Lagrangian methods could be set equal to μ^e, which is the velocity-dependent electromagnetic mass of the electron.

However, Poincaré demonstrated in his classical papers of (1905) and (1906) that the electromagnetic world-picture as originally conceived cannot be carried through in a Lorentz covariant manner. The reason is that the energy and momentum of a particle, with mass deduced from the Lagrangian in Eq. (3.24), do not transform properly under Lorentz transformation. The supplementary forces necessary to rectify this problem serve also to hold Lorentz's deformable electron together so that it does not explode from the enormous Coulomb repulsions between its constituent parts. (Abraham's theory of a rigid sphere electron could not be made Lorentz covariant. For a detailed historical account of the electromagnetic world-picture see Miller, 1981.)

For Abraham, Lorentz and Poincaré there was no self-energy problem because they assumed a finite radius for the electron. But such is not the case in quantum electrodynamics, which deals with point electrons because the concept of electron radius is ambiguous in quantum mechanics and also spoils Lorentz invariance. As Heitler (1936) put it in the first edition of his book, which is essentially a handbook for calculations, 'In the present state of the theory [which deals with point charges] the only way to proceed is to *ignore this self energy entirely*' (emphasis in original).

Let us further compare the classical theory of the electron with quantum electrodynamics. For this purpose we assume that the electron can be represented by a charged oscillator. Then from Eq. (3.33) the expansion in Eq. (3.29) can be expressed in terms of the parameter

$$\left(\frac{r_0}{\lambda}\right) \ll 1, \tag{3.34}$$

where λ is the wavelength characteristic of the force exerted on the oscillator. Thus, besides being small, the higher-order terms in the expansion of Eq. (3.29) depend on the structure of the electron, that is r_0, and so vanish in the limit of zero electron radius. From the correspondence principle one would expect that these higher-order terms are also small in quantum theory. But this is not the case because the contribution to the electron's self-energy from second-order perturbation theory diverges. From Eq. (3.34) the cutoff for divergent integrals is

$$h\nu \ll \frac{mc^2}{\alpha} = 137mc^2. \tag{3.35}$$

For energies beyond $137mc^2$ quantum electrodynamics was expected to be invalid because here radiation-reaction forces dominate for which there are no correspondence-limit analogues.[16]

3.4 From negative energy states to positrons

As a result of Waller's published realization in (1930a) that inclusion of the negative energy states is required to describe the scattering of X-rays by atoms and for achieving the classical limit of Thomson scattering, and then Hermann Weyl's conjecture (1929) that the negative energy electron is a proton, Dirac returned to the nature of negative energy states. (Dirac, 1930a, cites a private communication from Waller about Waller's results in Waller, 1930a.) From Dirac's equation it follows that the negative energy electron moves in an external field as if it had positive charge and positive energy. But when an arbitrarily varying perturbation is present there is no definite separation between positive and negative energy states. Even in the absence of an electromagnetic field, an electron can change from positive to negative energies with the amount of energy of at least $2mc^2$ 'being simultaneously emitted in the form of radiation'.

Making the association, like Weyl, of the negative energy electron with the proton raised several difficulties, continued Dirac (1930a); chief among them was nonconservation of electric charge when the positive energy electron becomes a negative energy electron. In order to cure this problem, as well as that of all positive energy electrons falling into the more stable negative energy states, Dirac proposed the concept of the vacuum, or 'normal state of electrification', as a sea of negative energy electrons in which all states with negative energies are occupied. The Pauli principle prevents other electrons from falling into these already occupied states. Only departures from exact uniformity can be measured. Vacant states in the vacuum are 'holes', that is protons. The possibility of charge nonconservation is cured thus: When a positive energy electron drops into a hole the electron and proton annihilate with emission of radiation. So, whereas in the original Dirac theory positive energy electrons dropped into negative energy states, in the hole version there is the possibility of annihilation and creation of an electron–proton pair. (To anticipate, Dirac would soon realize that the negative energy electron was not a proton but a positively charged electron with positive energy, a positron.)

The invention of holes changed the character of Dirac's theory from a single-particle into a many-body theory. For example, continued Dirac, in the scattering of light by an electron (free or bound) two different sorts of intermediate states can occur (energy is conserved only between initial and final states, but not in the intermediate states which are virtual): (i) a negative energy electron can absorb the incident photon and be raised to a positive energy electron, while the incident electron fills up the vacant hole with emission of the final photon; (ii) a negative energy electron emits the final state photon and becomes a positive energy electron, while the initial positive energy electron fills the hole and absorbs the initial photon. Consequently transitions to negative energy states are replaced by creation and annihilation of electron–proton pairs.

What about the inequality of electron and proton masses? This is an 'unsolved difficulty' wrote Dirac in a sequel paper (Dirac, 1930b) where he calculated the process $e^- + p \rightarrow 2\gamma$. He used for the mass in the Dirac equation the mean of electron and proton masses.[17]

Dirac then broached a point that would become central in the ensuing years, namely what to do with the infinite density of electricity from the vacuum which makes an infinite contribution to the Maxwell equation

$$\operatorname{div} E = -4\pi\rho. \tag{3.36}$$

Dirac proposed that the finite measured charge density ρ should be interpreted 'as the departure from the normal state of electrification in the world'.

Oppenheimer (1930a) went on to show that the transition probability for annihilation of an electron by a proton is 'absurdly large' and leads to an 'absurdly short mean life for matter' of 10^{-9} s.

In light of Oppenheimer's result and Weyl's proof in (1931) that negative and positive energy electrons must have the same mass, Dirac (1931) rethought the concept of holes. In a tone reminiscent of that of Hermann Minkowski in 1908, Dirac wrote that branches of mathematics thought to be 'purely fictions of the mind' have turned out to be essential for a 'description of the physical world', for example, non-Euclidean geometries and noncommutative algebra. The key problems of that time in theoretical physics, such as the relativistic formulation of quantum mechanics, would probably require 'a more drastic revision of our fundamental concepts than any that have gone before'. Most likely, he thought, the best way to advance was to apply the 'resources of pure mathematics' to perfect and generalize the existing formulation of theoretical physics and then to 'interpret the new mathematical features in terms of physical entities'. With deliberate underemphasis, Dirac excluded the word 'new' before 'physical entities'.

In (1931) Dirac pointed out that his 'Theory of electrons and protons' (1930a) had been 'a small step in this direction'. But now he offered a new physical entity, antimatter. Dirac proposed that 'a hole, if there were one, would be a new kind of particle, unknown to experimental physics, having the same mass and opposite charge to an electron'. As to protons, they 'have their own negative energy states with holes that are antiprotons' (see also Kragh, 1989).

4 Quantum electrodynamics

Meanwhile, on 18 September 1931, Heisenberg wrote to Bohr that 'my work seems to be somewhat gray on gray' (Pauli, 1985). By 1931 the bright future for quantum physics that seemed just over the horizon in the fall of 1927 paled with the formulation of relativistic quantum mechanics. Besides the as yet uninterpreted negative energy states, there was the electron's divergent self-energy and the continuous energy spectrum of β-particles in the supposedly two-body final state of nuclear β-decay, which implied that energy was not conserved in nuclear reactions (Bohr, 1932) and that perhaps quantum mechanics was not valid within the nucleus (to be discussed in Chapter 5).

4.1 Measurement problems in a quantum theory of the electromagnetic field

At the 20–25 October 1930 Solvay Conference, Bohr, Dirac, Heisenberg and Pauli concurred that fundamental difficulties in quantum electrodynamics might be clarified by investigating measurability of electromagnetic field quantities. Upon his return to Copenhagen from Brussels, Bohr continued discussing field measurements with Lev Landau, who happened to be visiting at Bohr's institute. In December 1930 Landau went on to Zürich, where he interested Pauli's assistant Rudolf Peierls in field measurements. Their deliberations led to a joint 1931 publication entitled 'Extension of the uncertainty principle to relativistic quantum theory' (Landau and Peierls, 1931).

Measurement of an electric field can be accomplished through measuring the change in momentum of a charged test body placed in the field. Landau and Peierls claimed that the change in momentum of a charged test body incurs an additional inaccuracy which enters through the radiation field it produces when changing momentum. This added inaccuracy depends on the charge of the probe and can exceed the uncertainty relations from quantum mechanics. Landau and Peierls concluded that electromagnetic field quantities cannot be measured in the quantum domain and assumed the root of the difficulty to be the negative energy states in Dirac's theory of the electron: '... it would be surprising if the formulation of [quantum electrodynamics] bore any resemblance to reality. [In] the correct relativistic quantum theory ... there will therefore be no physical quantities and no measurements in the sense of wave mechanics.'

But in Landau and Peierls' view the most insidious problem in quantum electrodynamics was the unmeasurability of the electron's position, which is an inherent part of the theory owing to negative energy states.[18] As further support for the radical view of overthrow of foundations, Landau and Peierls offered the continuous energy spectrum of β-rays in nuclear β-decay. Landau and Peierls' results bore not only on the validity of quantum electrodynamics, but more generally on how a theory of submicroscopic phenomena ought to be structured. Fundamental measurement problems that everyone assumed to have been settled in 1927 surfaced again.

A flurry of letters were exchanged between Bohr, Heisenberg and Pauli. Eventually Bohr himself acted to correct the situation, as he wrote to Pauli on 25 January 1933 (Pauli, 1985):

> [W]e surely all agree that we stand before a new developmental phase, one that requires new methods. What to me stands at the heart of the situation is that we should not anticipate this development through misuse of apparently logical arguments. The necessary caution to the danger of the Landau and Peierls mentality has indeed led to a closer investigation of the limits of the quantum electrodynamical formalism. Even if I could not find an error in Landau and Peierls' arguments on the measurement of electromagnetic field quantities, their criticism of the formalism's foundation always made me uncomfortable [since] the particle problem does not enter explicitly. Together with Rosenfeld I have in the autumn undertaken the task and we have arrived at the result that full agreement stands between the principal limits of the measurements of the electromagnetic force and the exchange relations of the field components in the formalism. We show that the measurement disturbance caused, according to Landau and Peierls, by the emitted radiation of the probe particle can be totally eliminated.

The results are in the 1933 paper 'On the question of the measurability of electromagnetic field quantities' (Bohr and Rosenfeld, 1933). Bohr and Rosenfeld's analysis reveals that optimal field measurements, that is minimum uncertainties, are made with macroscopic field probes with the largest possible charge density and so whose atomic constitution need not be considered. As Pauli put it in a letter to Peierls of 22 May 1933, which includes as well a lesson in the meaning of the uncertainty relations:

> If a measurement inaccuracy-inequality exists that does not emerge directly from the theoretical formalism as an exchange relation, then one must have assignable physical grounds for one such relation. For the relations in your work these grounds are the factual errors due to the negative energy states . . . . But for the field measurement the negative energy states play no role [because the exchange relations in Eq. (2.57)–(2.59)] do not contain the charge, mass, or dimensions of the probe body.

Besides demonstrating inadequacies and errors in Landau and Peierls' conclusions, Bohr and Rosenfeld deduced new complementarities in a relativistic treatment of electromagnetic fields, which they referred to as 'quantum field theory'. Their analysis further elucidated the measurement situation in nonrelativistic quantum mechanics.[19]

The paper made an enormous impact on Bohr. As he wrote to Pauli (1985) on 15 February 1934, 'eliminating electrons with their divergent self-energy from measurements of the electromagnetic field. . . was for me a great liberation'. That the 'field concept' must be applied with caution could well reside in the fact that 'all field actions can be observed only through their effects on matter'. Bohr began to look at fundamental problems differently. For example, since we can consider classical electron theory to be an idealization valid for actions greater than h, then why not consider quantum electrodynamics to be an idealization whose concepts are valid for charges much larger than the fundamental charge? Then there ought to be a correspondence-limit argument for taking quantum electrodynamics down into smaller space-time regions. Evidence for such a correspondence principle is 'Einstein's 1916 derivation from essentially correspondence principle arguments [from the old Bohr atomic theory] of the radiation law that has as one of its consequences fluctuation phenomena'. Bohr suggested further investigations of field quantization and measurement theory to be the route to a consistent theory of matter free of divergences. This path 'would make research purely illusory toward further construction of hole theory'. As a start Bohr proposed (with no details given) the application of a version of quantum electrodynamics and measurement theory to atomic physics problems.

On 23 February 1934 Pauli wrote to Heisenberg that Bohr's statements on the application of field theory and measurement theory to atomic physics were '*ganz konfus*' (Pauli, 1985). Nevertheless, by late 1934, when the situation itself in quantum electrodynamics had become *ganz konfus*, Heisenberg would return to a version of Bohr's proposal to fold measurement theory into quantum electrodynamics (Section 4.11).

4.2 Heisenberg's first attempt at a fundamental length

An early attempt by Heisenberg to eliminate both the electron's self-energy and negative energy transitions was a 1930 proposal to chop up space-time into cells of volume l^3, where $l = h/Mc$ is the 'fundamental length', and M is the proton mass. Heisenberg sketched this 'lattice world [*Gitterwelt*]' in a letter to Bohr of March 1930 (Pauli, 1985).

Considering in some detail only a one-dimensional case, Heisenberg used a difference equation limit of the one-dimensional Klein–Gordon equation, which is

$$\left(\frac{\mathrm{d}^2}{\mathrm{d}x^2} - \frac{1}{c^2}\frac{\mathrm{d}^2}{\mathrm{d}t^2} - \frac{m^2c^2}{\hbar^2}\right)\psi = 0. \tag{4.1}$$

From Eq. (4.1) Heisenberg obtained

$$\left(\frac{E}{c}\right)^2 u_n - \frac{\hbar}{il}(u_{n+1} - 2u_n + u_{n-1}) - \frac{m^2c^2}{\hbar^2}u_n = 0. \tag{4.2}$$

Eq. (4.2) is the analogue for oscillations of a linear chain of masspoints whose energy is given as a function of the quantum number n. The solution is a periodic function with maxima and minima at $(2n + 1)\pi$ and $2n\pi$, respectively. In the neighborhood of a minimum an electron behaves like a normal electron, and in the neighborhood of a maximum it behaves like a proton. (Dirac had not yet postulated the positron.) Extension to three dimensions proved disasterous: energy, momentum and charge conservation were violated.

Heisenberg discussed the lattice world in 'The self-energy of the electron' [1]. According to the correspondence limit with classical electromagnetic theory, the quantum self-energy of a point electron is likewise infinite. But a finite radius electron is unacceptable in quantum mechanics. A lattice world prevents this problem, whilst introducing others such as lack of relativistic invariance due to a fundamental length.

In order to demonstrate that the causes of divergences in field theory reside also in its 'foundation', Heisenberg dropped the lattice world and considered instead electrons of high energy $E \gg mc^2$, so that he could set $m = M = 0$ in a c-number, that is single-particle theory of an electron interacting with its own electromagnetic field. In this case Heisenberg noted that only the constants h, c and e appear, from which a length cannot be constructed.

The seriousness of a divergent self-energy is underscored by Heisenberg's statement on whether 'considerations in terms of the correspondence principle can be no longer useful' (recall the letter of Pauli to Peierls, 18 June 1929, quoted from in Section 3.2). Whereas in pre-relativistic theories of the electron, the electron's energy is given entirely by

$$H = \frac{1}{8\pi}\int(\boldsymbol{E}^2 + \boldsymbol{H}^2)\,\mathrm{d}V$$

(see Section 3.3), in quantum theory the interaction between matter and field (see Eq. (3.22)) must be considered. Moreover, continued Heisenberg, in the Dirac theory of the electron $\mathrm{d}q_i/\mathrm{d}t = c\alpha_i$, and so an electron seems to move at the velocity of light (see note 18).

An essential ingredient of Heisenberg's calculations was whether the zero-point vacuum energy could be eliminated without use of 'formal artifices', which are defined through relations of the sort

$$-\Delta^{1/2} \exp ikx = |k| \exp ikx \tag{4.3}$$

$$\frac{1}{-\Delta^{1/2}} \exp ikx = \frac{1}{|k|} \exp ikx \tag{4.4}$$

that permit the field Hamiltonian operator to be written as

$$H = \frac{1}{8\pi} \int \left\{ (\boldsymbol{E}^2 + \boldsymbol{H}^2) + \mathrm{i} \left[E \frac{1}{-\Delta^{1/2}} \operatorname{curl} H \right] \right\} \mathrm{d}V. \tag{4.5}$$

The second term in Eq. (4.5) serves to eliminate the zero-point energy. In a letter to Pauli of 21 July 1933, Heisenberg referred to the 'horrible [*abscheulich*] operator $1/\Delta^{1/2}$ which we cannot view as *anschaulich*' (Pauli, 1985). Pauli recalled Heisenberg as saying that 'I categorically refuse to calculate with the $\Delta^{1/2}$-operator' (letter of Pauli to Ehrenfest of 28 October 1932, in Pauli, 1985). Most likely Heisenberg's disgust at the $\Delta^{1/2}$ functions resided in their nonlocal behavior and consequent improper Lorentz transformation properties.[20]

4.3 An 'intuitive' time-dependent perturbation theory

Almost on the eve of jumping off into nuclear physics, Heisenberg formulated a method for applying time-dependent perturbation theory to the interaction between a semiclassical radiation field and a q-number formulation of Dirac's theory of the electron with a wave function

$$\Psi_\sigma(r, t) = \sum_n a_n u_n(r, \sigma) \exp(-2\pi \mathrm{i} E_n t/h), \tag{4.6}$$

where the index σ denotes the electron's spin and $a^\dagger_n a_n$ represents 'intuitively' the number of electrons in the state n, or also the intensity of the proper vibration u_n, and has eigenvalues 0 and 1. By 'semi-classical radiation theory' Heisenberg meant that electric and magnetic fields are expressed as waves with coefficients b and $b^\dagger$ that are q-numbers (see Eqs (3) and (4) in [2]). Heisenberg calculated the operator a_n from time-dependent perturbation theory applied to the Dirac equation (see Eq. (3.13)).

Heisenberg's reason for formulating this calculational method was to connect quantum electrodynamics more closely than before with 'intuitive conceptions of classical theory and of wave mechanics', thereby with the correspondence principle too. In order to 'escape the well-known difficulties of quantum electrodynamics (infinite self-energy)' he calculated a_n and $a^\dagger_n$ only to first order.

For Heisenberg a bonus of this method was its demonstration that Schrödinger's 'intuitive idea' that Eq. (22)–[2] generated radiation can be maintained only if the a_n's are operators.

4.4 Multiple-time theory, hole theory and second quantization

Perhaps depressed about the current state of his theory, Dirac turned to formulating a new version of quantum electrodynamics. As a guide he relied upon correspondence-limit procedures. Dirac (1932) advocated a return to Heisenberg's (1925) method of formulating a quantum mechanics in terms only of observable quantities. For relativistic quantum mechanics Dirac suggested that the important quantities ought to be the 'various monochromatic components of the ingoing and of the outgoing fields of radiation' from which probability amplitudes are formed. (Similar quantities would make their appearance in Heisenberg's 1943 S-matrix theory as the in and out states.) According to Dirac this formulation was an improvement on the Heisenberg–Pauli quantum electrodynamics in which field and particle were treated as dynamical systems on the same footing, which is not satisfactory because the electromagnetic field is also used to make measurements and therefore 'should appear in the theory as something more elementary and fundamental'. In order to render the new theory manifestly covariant, Dirac proposed that each interacting electron should have its own time variable.

Pauli's comment to Dirac on the new formulation of quantum electrodynamics was that it is 'certainly no masterpiece' (letter of Pauli to Dirac, 11 September 1932, in Pauli, 1985).

Using what became known as the interaction representation, Léon Rosenfeld proved the equivalence of Dirac's new theory with the Heisenberg–Pauli quantum electrodynamics (Rosenfeld, 1932).

Dirac, Vladimir Fock and Boris Podolsky (1932) elaborated on Rosenfeld's proof, which they considered to be 'obscure and does not bring out some features of the relation of the two theories'. Bloch (1934) went on to demonstrate that wave functions in a many-time formalism 'can be interpreted analogously to the wave functions of ordinary wave mechanics as a probability amplitude for such measurements as are made at times t_s at the particles s and at the time t on the electromagnetic field', where t is the 'field time' and t_s is the 'particle time' and $s = 1, 2, \ldots$, n, for each of the n particles.

Despite further work done on the multiple-time formalism, it dropped into obscurity once its equivalence with the Heisenberg–Pauli quantum electrodynamics was proven rigorously by Dirac, Fock and Podolsky.[21]

During 1932, Heisenberg was hard at work inventing nuclear physics (see Chapter 5). Pauli occupied himself with more formal matters like general relativity and starting his *Handbuch* article. In correspondence Pauli continued to discuss quantum electrodynamics, for example the zero-point energy, and to critique Dirac's negative energy states.

To Ehrenfest, Pauli wrote (28 October 1932, Pauli 1985) that in order to eliminate the negative energy states a future theory would require 'modification of the space-time concept (not only the field concept) in regions of the dimension of h/mc, or, h/mc^2'. In this letter Pauli, to the best of my knowledge, noted for the first time that the 'light quantum number N and $\boldsymbol{E}$ or $\boldsymbol{H}$ do not commute, so experimental arrangements for measurement of these quantities are mutually exclusive (complementarity like p and q)'.

In 1933 Heisenberg and Pauli resumed their correspondence. On 18 January 1933, Pauli (1985) expressed a distaste for second quantization:

> By 'second [*nochmaliger*]' quantization I mean the following. The transition from *classical* particle mechanics to the c-number waves Ψ in three-dimensional space I call quantization. The transition from the c-number Ψ waves to the q-number Ψ waves I also call quantization; so: 'second' quantization. These designations pursue the deliberate aim that the reader *ought to* find disgusting! – I find especially repugnant (naturally this is not meant as an objection in the *logical* sense), that in the case of the exclusion principle states must be brought into a definite order of succession, that is, defined with the Jordan–Wigner α_n. [Emphasis in original.]

The factor α_n enters through the anticommutation properties of the creation and annihilation operators for electrons and determines whether a minus or plus sign is inserted before an ordering of electron states, depending on how many occupied states precede the nth state.

Pauli continued in this letter: 'Despite my momentary aversion for the q-number field for matter, I consider it not impossible that in a future theory of the many-body problem such mathematics can be of use. But for what, we know not today.'

On 19 April 1933 Pauli wrote (in English) to P. M. Blackett (Pauli, 1985): 'I don't believe on [*sic*] the Dirac-"holes", even if the positive electron exist [*sic*]'.

And to Dirac (in English) on 1 May 1933 (Pauli, 1985): 'I do not believe on [*sic*] your perception of "holes", even if the existence of the, "antielectron" is proved.' Pauli expressed a similar pessimistic attitude in his *Handbuch* article.

Pauli's statements about the positive electron were made in the face of empirical evidence for their existence which had first appeared in Carl Anderson's 1931 paper. By 22 May 1933 Pauli accepted the new particle, as the next letter attested, but with an unusual interpretation.

In a letter to Peierls of 22 May 1933 Pauli reemphasized his belief in the power of the conservation laws of energy, momentum, angular momentum and in Bose and Fermi–Dirac statistics. For example, that the newly discovered neutron obeys Fermi statistics accounts for the nonoccurrence of a hydrogen atom becoming a neutron. Nevertheless, Pauli was willing to support an idea proposed by Walter Elsasser that the positron is a Bose particle with spin zero or one (Pauli, 1985): 'The fact that this viewpoint is opposed to the Dirac hole theory speaks only *for* it' (emphasis in original). As further support for Elsasser's proposal Pauli pointed out that 'as experience shows positive and negative electricity are not in the same ratio and it is very unattractive to me to establish this asymmetry in the initial state [primitive state] of the world as Dirac does'.

Heisenberg opposed Pauli's view: 'In nature, between the two processes: electron emission and positron emission, seems to reign complete symmetry. But naturally we must be resigned to great surprises' (letter of Heisenberg to Arnold Sommerfeld, 17 June 1933, in Pauli, 1985).

In a letter to Heisenberg of 24 July 1933, Pauli commented on the nature of Heisenberg's *Austausch* and the positron. (Heisenberg's *Austausch* or exchange force is discussed in Chapter 5.) If the neutron has spin $\frac{1}{2}$ then it cannot decay into a proton and electron. The neutrino helps here because the process $n \rightarrow p + e^- + \nu$ satisfies conservation of charge and angular momentum. Pauli speculated the inverse reaction to be $p \rightarrow n + e^+$, 'which is possible only if the positive electron satisfies Bose-statistics and has integral number spin (Elsasser)' (Pauli, 1985). In the hole theory, continued Pauli, there must be complete symmetry between positive and negative electrons and so the analysis becomes complicated for neutron and proton decay. If hole theory is correct, continued Pauli, then Heisenberg's '*Austauschkraft* between neutron and proton does not exist'.

Concerning hole theory in general, Pauli emphasized, as he would elsewhere, that the

> play with infinite concepts in the present framework is unacceptable and will eventually lead to contradictions. In particular, I don't see how the Coulomb interaction energy can be eliminated unequivocally and free from arbitrariness with the infinitely many occupied states [that is, Dirac's description of the vacuum as the state where all negative energy states are occupied].

Pauli believed that there would be no infinite quantities in a version of quantum electrodynamics without any negative energy states which maintained gauge and relativistic invariance, and in which there was pair production.

In a letter to Pauli of 17 July 1933, Heisenberg described an attempt to fold hole theory into a second-quantized version of quantum electrodynamics (Pauli, 1985). This work draws on the second-quantized Hamiltonian from their

1929 opus for an electron interacting with an electromagnetic field, which is

$$H = \sum_{E>0} N_s E_s + \sum_{E<0} N_s E_s + \sum_k N_k h\nu_k + H_{\text{int}}, \tag{4.7}$$

where $H_{\text{int}} = -ie\bar{\psi}\gamma_\mu \psi A^\mu$; the summation for the electrons includes positive and negative energies; and the term $\sum_k N_k h\nu_k$ is from the quantized electromagnetic field (Heisenberg excluded the zero-point energy). Heisenberg replaced the occupation number N_s in the negative energy term by $1 - N_s'$, and so the electron term becomes

$$H' = \sum_{E>0} N_s E_s + \sum_{E<0} E_s(1 - N_s'), \tag{4.8}$$

with similar rearrangements in the interaction term.

The infinite negative charge of the Dirac sea is contained in the term $\sum_{E<0} E_s$, which Heisenberg proposed to omit provided that the interaction is reordered so that all creation operators stand to the left of all annihilation operators. Although Heisenberg had applied the Klein–Jordan trick, he noted that this rewriting of the Hamiltonian does not get rid of the electron's electrostatic (Coulomb) self-energy. Nevertheless,

> I believe that the swindle [of dropping $\sum_{E<0} E_s$ in the resulting Hamiltonian] is not as fatal as any other swindle of quantum electrodynamics (self-[energy]) and scheme 3 [Eq. (4.8) plus $H_{\text{interaction}}$ with properly rearranged creation and annihilation operators] seems to me as equally solidly (or at least as solidly) founded, as the entire quantum electrodynamics (Pauli, 1985).

The gist of Pauli's 19 July 1933 detailed critique of Heisenberg's new 'scheme 3' is that since the self-energy problem remained then Heisenberg's scheme could well lead to other 'absurd consequences' (Pauli, 1985). Pauli expressed his preference for a coordinate space representation for quantum electrodynamics rather than having to deal with the requisite sign ordering functions for electrons in second quantization.[22]

Heisenberg replied (21 July 1933) that he 'does not believe that a formulation of a h[ole] th[eory] of particles in coordinate space is possible – at best it would be artificial and unsatisfying' (Pauli, 1985). For example, according to 'scheme 3' the Maxwell equation for the charge in configuration space is

$$\text{div} E = -4\pi e(\psi^{(+)*}\psi^+ + \psi^{(-)*}\psi^+ + \psi^{(-)*}\psi^- - \psi^{(+)*}\psi^-), \tag{4.9}$$

where the fields in configuration space are

$$\psi = \psi^+ + \psi^-, \tag{4.10}$$

$$\psi^* = \psi^{(+)*} + \psi^{(-)*}, \tag{4.11}$$

ψ^+ (ψ^-) is the wave function for an electron (positron), and Heisenberg maintained the minus sign in Eq. (4.9) from a reordering of terms in occupation number space because the vacuum expectation value of the charge density diverges if ρ is written as Eq. (3.8). Contrary to hitherto unpublished work by Peierls and Pauli (here Heisenberg referred to work probably discussed in person), one cannot use the separate fields in anticommutation relations like

$$\{\psi^{+}(x), \psi^{(-)*}(x')\} = \delta(x - x'), \tag{4.12}$$

but only the full field

$$\{\psi^{(-)*}(x) + \psi^{+}(x), \psi^{-}(x') + \psi^{(+)*}(x')\} = \delta(x - x'). \tag{4.13}$$

Consequently, wrote Heisenberg, 'on these grounds I see no possibility to introduce in any simple way the probability to find a positive el[ectron] at the position x'. For example, to measure with a γ-ray microscope whether at the time $t = t_0$ an electron is at the position $x = x_0$ to within a Compton wavelength h/mc would lead to pair creation. Only for light of energy $h\nu \ll 2mc^2$ does this sort of experiment make sense. The hole theory 'will lead to various horrors' as long as the self-energy problem was not resolved, wrote Heisenberg.

On 22 July 1933 Pauli replied to Heisenberg that Peierls had found that variation of the electric charge density by an external potential leads to convergence problems just like the self-energy. This is an early intimation of vacuum polarization. Pauli suggested to Peierls that there was a cutoff at the classical electron radius for all divergent integrals (Pauli, 1985).

On 25 July 1933 Pauli replied to Heisenberg's letter of 21 July 1933. Pauli agreed that there were difficulties with a spatial probability density but 'grants in no way that this difficulty is in the least altered or removed when we introduce as variables [occupation numbers N] instead of spatial position' (Pauli, 1985). Attempting to measure how many particles are in a definite state leads to production and/or annihilation of pairs. In Pauli's view 'there are no satisfying formulations of the hole idea, only different but mathematically equivalent, although more or less aesthetically appearing formulas, with the same physical difficulties'.

Pauli wrote that, two years before, Peierls and he had found that the 'most aesthetically appearing formulation' for particles is in the configuration space of the momentum coordinate where one need not deal with 'sign functions and occupation numbers. Then we could Fourier transform to coordinate space and discuss measurements and a probability interpretation. [But] I believe that the description of nature by wave functions is not at all the ultimate end of all knowledge'.

On 29 September 1933 Pauli wrote to Heisenberg that like Bohr and Heisenberg, he was 'neither approving nor disapproving of hole theory' (Pauli, 1985). As grounds for not *a priori* rejecting hole theory, Pauli cited its predictions of the positron, pair production and the correct calculation of pair production by Oppenheimer and Plessett (1933), which turned out to be incorrect.[23]

4.5 Dirac at Solvay in 1933: vacuum polarization

In September 1933 at Solvay [3], Dirac reiterated his proposal that the vacuum state is the one where all positive energy states are empty and all negative energy states are filled. He then turned to formulating a method to calculate the measured finite charge density in the Maxwell equation div $E = -4\pi\rho$ by subtracting the infinite charge density of the Dirac sea.

It is instructive to use second quantization to see how this occurs, although, of course, the result is the same in configuration space in which all calculations are done in [3]. Dirac's definition of the vacuum is:

$$N_m = \begin{cases} 1 & E_m < 0 \\ 0 & E_m > 0, \end{cases} \tag{4.14}$$

where $N_m = b^{\dagger}_m b_m$ is the occupation number for the electron state m. In the absence of external fields the Hamiltonian and total charge operators

$$H = \sum_{E_m>0} E_m N_m + \sum_{E_m<0} E_m N_m \tag{4.15}$$

$$Q = \sum_{E_m>0} N_m + \sum_{E_m<0} N_m \tag{4.16}$$

are not zero for the vacuum, but $-\infty$. Term-by-term subtractions from Eqs. (4.15) and (4.16) by the infinite quantities $\sum_{E_m<0} E_m$ and $e\sum_{E_m<0} \mathbf{1}$, respectively, yield

$$H = \sum_{E_m>0} E_m N_m + \sum_{E_m<0} |E_m|(1 - N_m) \tag{4.17}$$

$$Q = \sum_{E_m>0} N_m - \sum_{E_m<0} (1 - N_m), \tag{4.18}$$

whose vacuum expectation values vanish, as a result of the 'subtraction between two infinities', [3].

When an external electromagnetic field is present, continued Dirac, matters are even more complicated because there is no clear distinction between positive and negative energy states and special subtraction rules must be given. At Solvay, however, Dirac was not yet ready to carry out this sort of subtraction program. Rather he used the Hartree self-consistent field approximation and symbolic operator methods to set up a perturbation calculation for the density matrix in Eq. (1)–[3]. The perturbation is a weak electrostatic field.

Dirac started the calculation in the momentum representation Eq. (8)–[3] and then transformed to the coordinate representation (Eq. (9)–[3]) in order to obtain the diagonal matrix element of the perturbed density matrix R_1, which to this order of approximation is the extra charge density induced by the external electrostatic field. In order to avoid the divergent term in Eq. (9)–[3], Dirac cut off the divergent integral at an electron energy corresponding to the electron's classical radius, within which quantum electrodynamics was assumed to be invalid. The result in Eq. (10)–[3] is the polarization of the vacuum by an external electrostatic field.

Using hole theory Dirac interpreted vacuum polarization as follows: An electron introduced into the vacuum disturbs the distribution of negative energy states owing to the action of the electron's own electrostatic field. The new distribution gives rise to a new extra 'density induced' in the vicinity of the electron that 'neutralizes a fraction' of the electron's 'true' charge to give the 'observed' charge. In modern quantum electrodynamics Dirac's 'true' charge is the bare charge, 'observed' charge is the one on the dressed electron, and neutralization of a fraction of the electron's charge is charge renormalization.

Dirac could have explained vacuum polarization using electron–positron concepts from second quantization thus: The electron's electrostatic field produces electron–positron pairs. Negative charge is repelled away from the electron, which becomes surrounded by a cloud of positrons that act to reduce the electron's 'true' charge to the 'observed' charge. This situation is analogous to insertion of a charged body into a polarizable medium. The effective or measured charge on the body is the original charge divided by the dielectric constant, and is less than the original charge. Consequently, the vacuum acts like a polarizable medium.

Dirac considered the first (divergent) term in Eq. (10)–[3] to be important, whereas the second term

$$-\frac{4}{15\pi}\frac{e^2}{\hbar c}\left(\frac{\hbar}{mc}\right)^2 \Delta\rho \tag{4.19}$$

becomes of interest only when the density ρ varies rapidly over distances of the order of the electron's Compton wavelength. This turned out not to be the case, as we discuss in a moment.

At Solvay Pauli finally found something useful in Dirac's hole theory. With wry understatement Pauli's comment on Dirac's lecture was [3]:

> The theory of holes always seemed very interesting to me on account of the role played in it by the exclusion principle. Whereas this principle was formerly only

> an isolated rule, of which the validity was independent of those of the bases of quantum theory, the theory of holes . . . would have been impossible if we had not wished to exclude all wave functions that are not antisymmetric. However, the general aspect of the theory is not satisfactory owing to the manner in which it is obliged to use the concept of infinity.

4.6 The Heisenberg–Pauli collaboration on positron theory

During 1933–4 Heisenberg and Pauli worked in close collaboration on positron theory. Despite Dirac's presentation of the density matrix at the Solvay Conference, they initially investigated elimination of divergences in positron theory along lines proposed in their 1929 Hamiltonian quantum electrodynamics formulation. Their density-matrix considerations began in mid-November 1933 after Pauli received a lengthy letter from Dirac describing the new formalism in more detail than at Solvay (11 November 1933, not in Pauli collection).

Their principal problem was to formulate procedures for subtracting light cone singularities from a relativistic version of Dirac's density matrix in a coordinate space representation for the case when not all the negative energy states are filled. One could then calculate the finite contributions from its diagonal matrix elements, which are the charge and current densities. Heisenberg, in particular, was interested in combining canonical field quantization from the Heisenberg–Pauli formulation of quantum electrodynamics with Dirac's density matrix. All of this had to be accomplished within a framework that was relativistically and gauge invariant, in agreement with the conservation laws of energy and momentum, and charge symmetric between electrons and positrons.

The 'Klein–Jordanschen Trick', as Pauli and Heisenberg playfully referred to it in mixed English and German, emerges throughout their correspondence. Whilst Pauli preferred to avoid it, in contrast, Heisenberg wrote to Bohr on 12 March 1934 (Pauli, 1985), 'I consider this "trick" of Klein and Jordan by no means to be a superficial artifice but its consistent utilization sets a definite formal path in the quantum theory of wave fields, which we must utilize in the future'.

On 25 January 1934 Heisenberg wrote an 'aesthetic remark' on the hole theory, namely that it should be 'invariant against the transformation $+e \rightarrow -e$' (Pauli, 1985).

With an intentional Bohrism, Pauli replied (27 January 1934) to Heisenberg that 'we must be prepared' to generalize Dirac's formalism (Pauli, 1985). Pauli hoped to soon receive from Dirac a more detailed density-matrix paper.

Heisenberg wrote on 27 January 1934 that the scheme they were considering at that time was 'unnecessarily complicated' (Pauli, 1985). He expressed his 'hole pessimism' on the program: There were complications in showing that expectation values of charge density and energy density are finite, and it is unclear whether in their version of hole theory the expectation value of the energy density will be negative.

Pauli agreed with Heisenberg's pessimism (letter of 30 January 1934, in Pauli, 1985) and set guidelines for their work:

'(I) Finite energy density
(II) Energy-momentum density finite and positive definite.

This can assure first the fixing of $e^2/\hbar c\ldots$'. Pauli suggested that initially they should leave the self-energy problem open and work toward a 'program' in which the interaction energy between matter and field is a perturbation in $e^2/\hbar c$. But Pauli was pessimistic that even (I) could be satisfied owing to the electron's divergent self-energy and incorrect results in pair production for photon energies $E \gg mc^2$.

Heisenberg wrote on 5 February 1934 that the root of the self-energy problem is the infinitely many degrees of freedom in the intermediate states (Pauli, 1985). Perhaps this infinity could be avoided in a theory with a finite number of degrees of freedom. But according to contemperaneous physics, this was not possible. For example, consider an electron and a positron in a box. After a while we expect that a Planck radiation distribution will arise with a temperature given by the energy of the initial pair. Consequently from two particles arbitrarily many can appear by production and annihilation. In Heisenberg's opinion the 'Klein–Jordan Trick' is the single most promising method to remove the self-energy. The key assumption involved in implementing the 'Klein–Jordan Trick' is to assume the existence 'so to speak of basic constituents of nature'. For given such particles (which are also solutions of the theory when no external forces are present – 'trivial solutions of the theory'), then interactions can be constructed of the form $\psi^+\chi^+\ldots\psi\chi$, where ψ and χ are second-quantized wave functions for electrons and positrons, and in all orderings creation operators stand to the left of annihilation operators. This ordering is accomplished by the 'Klein–Jordan Trick'.

In this scheme Heisenberg proposed that a correct wave equation would be somewhat like the following:

$$H\psi = \left\{ \sum a_k^\dagger a_k + \sum h\nu_\sigma b_\sigma^\dagger b_\sigma + \sum f_{k\sigma lmn} b_k^\dagger a_\sigma^\dagger a_l^\dagger a_m a_n \right.$$
$$\left. + \sum f_{k\sigma lmn} a_n^\dagger a_m^\dagger a_\sigma b_k \right\} \psi, \tag{4.20}$$

where the f's are cross-sections, that is 'measurable quantities', and a (b) is the annihilation operator for an electron (positron). Heisenberg next asked the question of whether it is possible to arrive at an equation like Eq. (4.20) without having 'to forfeit correspondence to the usual theory'. He went on to make a proposal which he 'has not followed in all details', namely to replace the usual wave equations $(H - E)\psi = 0$ and $(G_k - I_k)\psi = 0$ (where G_k is the momentum) with a quadratic wave equation

$$\left(H^2 - \sum G_k^2 - \text{constant}\right)\psi = 0. \tag{4.21}$$

Heisenberg expressed doubts over Dirac's arguments against a quadratic wave equation in the absence of external forces. Rather, a quadratic wave equation permits 'transposition of factors'; presumably reference is made to the Klein–Jordan trick. Heisenberg continued that Eq. (4.21) should yield Eq. (4.20), which, in turn, ought to give the electron and photon as trivial solutions without any divergent self-energy: 'Consequently, it seems certain to me that we obtain in this manner a correspondence-limit correct theory without self-energy.'

Pauli replied on 6 February 1934 that he had received the manuscript for Dirac's 1934 paper [4] (Pauli, 1985). Up to that point through 'complicated calculations', some of which involved redoing Dirac's calculations, the conservation laws of charge and current were satisfied:

> Whether also the energy law is satisfied God only knows! . . . Thus is Dirac's law of nature set upon Mount Sinai. All is expressed mathematically very elegantly. But physically I am not at all convinced. What use is it for vacuum polarization to be finite, if the self-energy is infinite and the frequency for pair production incorrect? Consequently I am presently very disgusted with the hole theory.

Pauli disagreed with Heisenberg's argument against the possibility of reducing the infinite number of degrees of freedom in the intermediate states to a finite number; perhaps Planck's radiation law was not valid for distances less than the electron's radius. Furthermore, he argued, one may have to modify electromagnetic field concepts using spinor theory – reference here was made to Louis de Broglie's neutrino theory of light in which a light quantum is constructed from two neutrinos. Pauli went on to inquire how Heisenberg's Eq. (4.21) could become four equations in order to describe a neutrino, and he signed off with 'Yours (drowning in Dirac's formulae)'.

On 7 February 1934 Pauli wrote to Heisenberg that the concept of charge density is more fundamental than particle number and 'perhaps the most fundamental of all of physics' (Pauli, 1985). Although Pauli considered Heisenberg's Eq. (4.21) as important, he could see no place in it for electric and magnetic fields, which, after all, must be there if the electron is to be a trivial solution. However, it is not entirely trivial since it carries at least an electrostatic field of infinite range. Moreover, the 'equation div$E = \rho$ and the corresponding equation for the current cannot be squared and added'.

So, concluded Pauli, however one tries to avoid the self-energy problem the concept of electric charge enters. The key problem is to 'fix $e^2/\hbar c$ and the "*Atomistik*" of the electric charge'.

In a letter to Pauli of 8 February 1934 Heisenberg lashed out at Dirac's hole theory as 'learned junk that no human being can accept'. He then discussed Pauli's recent work on the density matrix from which, Heisenberg speculated, would arise a Hamiltonian of the sort in Eq. (4.20) (which turned out not to be gauge invariant; see the letter of Pauli to Heisenberg, 17 February 1934, in Pauli, 1985).

The important advance that Heisenberg made at this point was to suggest writing Eqs. (4.9)–(4.11) as q-number equations and then to replace the creation (destruction) operator for an electron (positron) with the destruction (creation) operator for a positron (electron), that is the first step toward a mathematical expression for charge conjugation. In this case Eqs. (4.10) and (4.11) are rewritten as q-number equations with

$$\psi = \psi^{(+)} + \psi^{(-)} \tag{4.22}$$

$$\psi^+ = \psi^{(+)+} + \psi^{(-)+}, \tag{4.23}$$

where now $\psi^{(+)}(\psi^{(+)+})$ annihilates an electron (positron) and $\psi^{(-)+}(\psi^{(-)})$ creates an electron (positron).

Heisenberg employed this new representation for the charge density operator, which is the density matrix in the limit where the initial and final momenta become equal:

$$\rho = e\sum\{a^\dagger_{r+}a_{s+}u^*_{r+}u_{s+} + a^\dagger_{r+}b^\dagger_{s-}u^*_{r+}u_{s-} + b_{r-}a_{s+}u^*_{r+}u_{s+} - b^\dagger_{r-}b_{s-}u_{r+}u^*_{s+}\}, \tag{4.24}$$

where + and − in the subscripts mean positive and negative energy states, r and s are spin indices, and the quantities u are free particle spinors from the Dirac equation. The key point in Eq. (4.24) is that its vacuum expectation value vanishes owing to inversion of the positron creation and annihilation operators. In contrast, the vacuum expectation value of ρ obtained directly from Eq. (3.8) diverges. In the terminology of modern quantum electrodynamics, Heisenberg had written the q-number charge density as a normal product (see, too, Section 2.3). In order to obtain this result without inverting factors requires writing the charge density operator in the form

$$\rho = -e\{\psi^+\psi - \langle 0|\psi^+\psi|0\rangle\}, \tag{4.25}$$

where the q-number wave functions in Eq. (4.25) are from Eqs. (4.22) and (4.23). The vacuum expectation value of the charge density operator in Eq. (4.25) vanishes owing to subtraction of the infinite quantity $\langle 0|\psi^+ \psi|0\rangle$. This is another example of obtaining a finite result from the ambiguous operation of 'subtraction between two infinities' and, to anticipate, this will be the procedure in Heisenberg's 'subtraction physics' in [6].

In the letter of 8 February 1934 to Pauli, Heisenberg maintained that Eq. (4.24) for the charge density 'takes care of itself', because it remains finite without any problems concerning how the limit of equating initial and final momenta is taken. Consequently Dirac's complicated calculations, some of which concerned the vacuum polarization, are '*völliger Luxus*'.

But all was not well with Heisenberg's ρ. In a letter of 10 February 1934 Pauli wrote that, in fact, ρ does not 'take care of itself' because in first approximation in a development in e, ρ becomes infinite 'totally analogous to Dirac's calculation at the Solvay Congress' (Pauli, 1985). Nevertheless, Pauli wondered if he perhaps made some sort of 'elementary error'.

Heisenberg replied (date unknown, but in February 1934) that Pauli was correct (Pauli, 1985). Whereas the total charge density in a certain volume remains finite in the presence of a perturbation, it can be undetermined at a point: 'this would not disturb me since the concept "charge density at a definite point" is as stupid as the concept "field at a definite point"'. Heisenberg speculated that this situation could be rectified in the same manner as one deals with the energy density in a certain volume, namely by 'smearing out the walls of the volume'.

Pauli suggested the method for Heisenberg to carry out his speculation, namely (letter of Pauli to Heisenberg, 13 February 1934, in Pauli, 1985): rather than 'smearing of the walls' one should consider averaging over space-time with the averaging method of Bohr and Rosenfeld:

$$\bar{e} = \bar{\bar{\rho}} = \frac{1}{V}\frac{1}{T}\int_V \mathrm{d}V \int_T \mathrm{d}t\, \rho(x,t). \tag{4.26}$$

Taking this hint, two days later Heisenberg reported to Pauli (letter of 15 February 1934, in Pauli, 1985) calculations on the fluctuations in charge owing to measurement over finite space-time intervals where pairs are produced due to the measurement apparatus. Heisenberg obtained convergent results and problems arise only for space-time intervals smaller than the electron's Compton wavelength and for well-defined (sharp) boundary walls. In these cases infinitely many pairs are produced (in the second case at the sharp boundary) (see Heisenberg, 1934b).[24]

In a letter of 17 February 1934 Pauli expressed tempered agreement with Heisenberg's results on charge fluctuations because he believed that basically all infinities that occur in hole theory were linked to the electron's infinite self-energy (Pauli, 1985). Toward curing this problem Pauli suggested a '*Dreimänner-Arbeit*' like in 1925, only now the frontier problems in physics would be cracked by Heisenberg, Pauli and Pauli's new assistant Viktor F. Weisskopf. Pauli included a detailed sketch of the proposed three-man paper 'Contribution to the theory of electrons and positrons'. Again Pauli emphasized that all difficulties with hole theory depended on fixing the value of the fine structure constant and solving the problem of the electron's self-energy. Basically, according to Pauli, the calculational problem that Dirac had only touched on was to find c-numbers that could be subtracted from the charge and current densities in order to render these quantities finite in a manner that is also gauge and relativistically invariant, like in Eq. (4.25).

On 26 February 1934 Pauli sent Heisenberg the manuscript of Dirac's (1934) paper (Pauli, 1985). Heisenberg then joined Pauli in dissecting Dirac's paper and improving the density-matrix formalism, particularly concerning its limit procedures for setting equal the initial and final momenta or coordinates, gauge invariance, energy conservation and charge symmetry. Succeeding letters sent between Heisenberg and Pauli indicate a shift in their previous work towards emphasis on Dirac's own formalism.

4.7 The subtraction physics

4.7.1 Dirac defines the problem

In [4] Dirac set the paper's goal thus:

> We now have a picture of the world in which there are an infinite number of negative-energy electrons . . . having energies extending continuously from $-mc^2$ to $-\infty$. The problem we have to consider is the way this infinity can be handled mathematically and the physical effects it produces. In particular, we must set up some assumptions for the production of the electromagnetic field by the electron distribution, which assumptions must be such that any finite change in the distribution produces a change in the field in agreement with Maxwell's equations, but such that the infinite field which would be required by Maxwell's equations from an infinite density of electrons is in some way cut out.
>
> These problems are quite simple when we suppose each electron to be moving in a space free of electromagnetic field. They are not so simple when there is a field present, since the positive- and negative-energy states then get mixed together so intimately that one cannot in general distinguish accurately between them in a relativistically invariant way.

Dirac defined the density matrix in a relativistic theory to be

$$\langle x't'|R|x''t''\rangle = \sum_{\text{occ}} \psi_{k'}(x', t')\psi^*_{k''}(x'', t''), \tag{4.27}$$

where the ψ's are four-component wave functions that are solutions of Dirac's equation for each individual electron, Hartree's method of the effective field is utilized,[25] k' and k'' are spin indices and (x', t') and (x'', t'') are two different space-time points. Eq. (4.27) can be rewritten as

$$\langle x't'|R|x''t''\rangle = \langle x't'|(R_F + R_1)|x''t''\rangle, \tag{4.28}$$

where

$$\langle x't'|R_F|x''t''\rangle = \tfrac{1}{2}\sum_{\text{occ}} \psi_{k'}(x', t')\psi^*_{k''}(x'', t'') + \tfrac{1}{2}\sum_{\text{unocc}} \psi_{k'}(x', t')\psi^*_{k''}(x'', t'') \tag{4.29}$$

represents the full distribution with all states occupied and becomes a δ-function in $x' - x''$ when $t' = t''$;

$$\langle x't'|R_1|\text{x}''\text{t}''\rangle = \tfrac{1}{2}\sum_{\text{occ}} \psi_{k'}(x', t')\psi^*_{k''}(x'', t'') - \tfrac{1}{2}\sum_{\text{unocc}} \psi_{k'}(x', t')\psi^*_{k''}(x'', t'') \tag{4.30}$$

represents deviations from the case where all negative energy states are occupied.

The charge density is the diagonal element of the density matrix summed over all spins

$$\sum_k \langle xt|\rho|xt\rangle_k = \sum_k \langle xt|R|xt\rangle_k \tag{4.31}$$

and is infinite owing to light cone singularities which require special limit prescriptions for passing from (x', t') to (x'', t'') – we recall that without proper subtractions in occupation number space the vacuum expectation value of the charge density diverges.

Dirac continued:

> The problem now presents itself of finding some natural way of removing infinities from $\sum_k \langle xt|R|xt\rangle_k$ and $\sum_k \langle xt|\alpha_s R|xt\rangle_k$ [which is the current density] so as to have finite remainders, which we could then assume to be the electric and current densities.

In an important calculation Dirac classified the light cone singularities of $\langle x't'|R_F|x''t''\rangle$ and $\langle x't'|R_1|x''t''\rangle$ for the case of no external field and claimed that with a field the singularities are of the 'same form . . . but have unknown coefficients'.

Consequently the density matrix can be divided into two parts

$$R = R_a + R_b \tag{4.32}$$

where R_a contains all singularities and '*the electric and current densities corresponding to R_b are those which are physically present, arising from the distribution of electrons and positrons*' (italics in original). The intent, of course, is to propose a counter-term $-R_a$ so that $R - R_a$ and, consequently, the measured charge densities are finite. This will be accomplished by Heisenberg in [6].

An immediate published response to Dirac's paper was by Peierls (1934), who at the time happened to be working with Dirac. Peierls questioned the uniqueness of Dirac's subtraction methods because the quantities involved are infinite. He succinctly put the goal of what Pauli would call the 'subtraction physics':

> In a recent paper Dirac proposes an expression for the current density which avoids the infinities occurring above [such as the vacuum expectation value of the charge density with subtraction terms]. Dirac does not look for a physical interpretation of [Eq. (4.25)], but starts from the postulate that the current density must be finite. [The 'recent paper' is our [4].]

Statements closer and closer to modern renormalization theory are beginning to appear.

4.7.2 *Weisskopf's calculation of the electron's self-energy in hole theory*

During March 1934 Viktor F. Weisskopf completed the first calculation in hole theory of the electron's self-energy [5] using Heisenberg's semiclassical representation of the electromagnetic field from 1931 [2] 'which is linked much more closely to classical electrodynamics' than the perturbation methods of Oppenheimer and Waller.

After separating an electron's electromagnetic field into curl free (electrostatic) and divergence free (electrodynamic) parts, to order e^2 the operators for the electrostatic and electrodynamic self-energies are

$$E^{\mathrm{S}} = \frac{1}{2}\iint \frac{[\rho(r) - \overline{\rho(r)}][\rho(r') - \overline{\rho(r')}]}{|r - r'|}\, \mathrm{d}r\, \mathrm{d}r' \tag{4.33}$$

$$E^{\mathrm{D}} = -\frac{1}{2}\int [i(r) - \overline{i(r)}][A(r') - \overline{A(r')}]\, \mathrm{d}r, \tag{4.34}$$

where $\overline{\rho(r)}$ and $\overline{i(r)}$ are the charge and current densities, respectively, of negative energy electrons, and $\overline{A(r)}$ is the vector potential that they generate. Whereas the operator quantities $\rho(r) - \overline{\rho(r)}$, $i(r) - \overline{i(r)}$ and $A(r') - \overline{A(r')}$ are constructed to have zero vacuum expectation values (by the 'subtraction physics'), the operators occurring in the numerator of Eqs. (4.33) and in (4.34) do not satisfy this requirement. Consequently Weisskopf had to subtract yet another infinity, namely the Coulomb and electrodynamical vacuum energies that he referred to as E^{S}_{vac} and E^{D}_{vac}, respectively. The self-energies calculated from Eqs. (4.33) and (4.34) he wrote as E^{S}_{vac+1} and E^{D}_{vac+1}, respectively, which refer to the energy of the vacuum plus one electron.

Consequently the electron's self-energies are

$$E^{S} = E^{S}_{vac+1} - E^{S}_{vac}, \tag{4.35}$$

$$E^{D} = E^{D}_{vac+1} - E^{D}_{vac}. \tag{4.36}$$

Weisskopf found that the divergence in E^{S} is logarithmic, an improvement over single-particle theory where it diverges linearly. But the leading divergence in E^{D} is quadratic, as in single-particle theory.

On 21 June 1934 Wendell Furry at the University of California, Berkeley, wrote to Weisskopf to advise him that 'about a year ago, at Prof. Bohr's suggestion, Dr. Carlson and I [recalculated E^{D} using hole theory and obtained a logarithmic divergence just like for the] electrostatic self energy'.[26]

In a 'correction' [5] Weisskopf explained the error. The new result showed that to order α (where $\alpha = e^2/\hbar c$ is the fine structure constant) the electron's self-energy is inversely proportional to its total energy and is logarithmically divergent.[27]

In a letter to Weisskopf of 11 February 1935, Heisenberg wrote that there may be other refinements to Weisskopf's calculation because on 'grounds of relativistic invariance we expect that

$$E^{S} + E^{D} = \text{const.}\, \frac{e^2}{h}(m^2c^2 + p^2)^{1/2} \int \frac{dk}{k}.\text{'}^{28} \tag{4.37}$$

It would turn out that in a covariant renormalized theory of quantum electrodynamics there is no need to break up the electron's self-energy into electrostatic and electrodynamic parts. Rather, the free electron's self-energy is logarithmically divergent and *inversely* dependent on its energy, which is the dependence that agrees with relativistic requirements (an argument contained implicitly in Weisskopf, 1939; see, too, Heitler, 1954).

4.7.3 *Beyond the correspondence principle*

Meanwhile, Heisenberg had other thoughts on what to do about the self-energy problem, which swirled about correspondence-limit procedures. For example, as he wrote to Bohr on 12 March 1934 (Pauli, 1985):

> My disposition regarding the problem of the infinite self-energy [is] somewhat different from what you have written about it. I am totally in agreement with

you that we possess presently no theory that goes essentially beyond application of the correspondence principle to Maxwell's theory, from which also follows that point charges must have an infinite self-energy – without reference to the formalism of quantum electrodynamics. But it seems to me that a reasonable quantization of fields goes beyond the correspondence principle and must remove the infinite self-energy in the same manner as Jordan and Klein have avoided the electrostatic part of the self-energy. I maintain that this 'trick' of Klein and Jordan is in no way a formal artifice, but on which the logical exploitation of a definite formal path in the quantum theory of wave fields must be exploited even more generally in the future. In the quantum theory of wave fields the problem of the *single* point charge appears analogous to consideration of a quantum mechanical system in the *deepest* quantum state; likewise since all 'downward' transition probabilities from the lowest quantum state of an oscillator must automatically vanish (which is a demand that does not follow from the correspondence principle but does not contradict it, however, and was important for the origins of quantum mechanics), we must promote in the quantum mechanics of waves that a single point charge is a trivial solution of the equations of motion, that is, that all transition probabilities vanish which give rise to the formation of another particle. This requirement does not contradict correspondence because the correspondence principle supplies only assertions on a system that contains many electrons or light quanta. Thus, I believe that if we conclude from correspondence principle arguments to the infinite self-energy (as, for example, Pauli and I must do in quantum electrodynamics) it is in reality just as false as if we want to maintain from correspondence grounds that the hydrogen atom in the ground state must radiate on account of the orbital motion of the electron. That it was hitherto not possible in quantum electrodynamics to remove the infinite self-energy, means truly that we are using an incorrect Hamiltonian function and we can probably first find the correct Hamiltonian function when e^2/hc is determined and when the hole theory is worked out. [Emphasis in original.]

In a letter to Pauli of 10 April 1934 Heisenberg calculated the charge density in density-matrix theory and obtained a result for which he claimed 'a polarization of the vacuum is no longer left over!' (Pauli, 1985). Even stronger, in a letter of 14 April 1934 to Pauli and Weisskopf, Heisenberg claimed that 'no polarization of the vacuum is necessarily correct and moreover trivial' (Pauli, 1985).

Pauli generally approved of Heisenberg's results, except that 'I believe not at all, that in your formalism the polarization of the vacuum emerges as *exactly* 0' (letter to Heisenberg of 17 April 1934, emphasis in original, Pauli, 1985). After this situation is corrected (how, Pauli knew not), 'your results, if correct, are truly the extreme that can be forced out of the hole theory. Everything will become *beautiful* when $e^2/\hbar c$ is fixed' (emphasis in original).

On 27 April 1934 Weisskopf informed Heisenberg that when subtractions were performed properly the polarization of the vacuum should be finite but not zero. Weisskopf wrote: 'Your result for the vanishing polarization was indeed suspicious on account of the existing pair production' (Pauli, 1985).

On 3 May 1934, through his intermediary Weisskopf, Pauli abruptly resigned from the density-matrix research program by mail in 'disgust'. Heisenberg went on to complete the work in June 1934 in his paper 'Remarks on the Dirac theory of the positron' [6], which can be considered as the closest thing to a renormalized quantum electrodynamics until 1948.

4.7.4 *Heisenberg's formulation of subtraction physics*

In [6] Heisenberg presented a version of Dirac's density-matrix formalism that is gauge and relativistically invariant and charge symmetric between electrons and positrons: 'Compared to Dirac's treatment, this paper emphasizes the significance of the conservation laws for the total system of radiation and matter and the necessity of formulating the basic equations in a manner extending beyond the Hartree approximation.' However, Heisenberg's demand at the beginning of the paper that the theory introduces 'no new infinities' is violated by the appearance of the photon's self-energy in the second-quantized version of density-matrix theory in Part II.

Part I is entitled 'Intuitive [*anschauliche*] theory of matter waves'. The term intuitive appears because the interaction between electrons is neglected and one has an 'intuitive correspondence-like picture of the actual process'.

Heisenberg emphasized that the important part of the density matrix R is R_F (Eq. (6)–[6] – the relation between Dirac's R_1 and Heisenberg's R_F is $R_F = R_1/2$) because R_F represents an imbalance between occupied states of positive and negative energies. Furthermore, R_F relates positive and negative charges in a manner in which charge symmetry achieves its highest form not in hole theory, but in second quantization, as Heisenberg showed in Part II.

Heisenberg motivated subtraction procedures for the density matrix with the matrix element for charge density (Eq. (9)–[6]), which is infinite even for the vacuum and so always requires subtraction, as in Eq. (4.25). For no external field present the vacuum expectation value of the charge density requires subtraction by a matrix that turns out to be uniquely determined by the electron distribution corresponding to a deviation from the situation where all negative energy states are occupied (Eqs (20)–(22) in [6]). When an external field is present Heisenberg made the claim that a uniquely determinable matrix can be found to be subtracted from R_F so as to remove its divergences.

In general (switching now to Heisenberg's notation), Heisenberg defined a new finite density matrix r as

$$r = R_S - S \tag{4.38}$$

where S is the matrix that cancels any infinities in R_S.

In [6] there are several trivial errors; among them is the sign in front of S. Another is that Heisenberg wrote the density matrix in his Eq. (1) incorrectly; it should be

$$\langle x't'k'|R|x''t''k''\rangle = \sum_n \psi(x't'k')\psi_n^*(x''t''k''). \tag{4.39}$$

For other errata see Heisenberg and Euler (1936).

Unlike the matrices R_S and R_F, r satisfies an inhomogeneous equation

$$Hr = HS \tag{4.40}$$

where S is a complicated function of the electromagnetic field and so 'is the natural expression of the fact that matter can be created and destroyed' (see Eqs (13) and (14) in [6]). But immediately there are difficulties with S because its free field value cannot be obtained uniquely by a simple limiting process from the case where fields are actually present: 'Taking into account the conservation laws of charge, energy and momentum permits restricting possibilities for S.'

Heisenberg went on to use time-dependent perturbation theory to calculate vacuum polarization in the 'intuitive theory of matter waves', that is hole theory. His method was as follows: The Dirac wave function for an electron in the presence of a static potential can be written as

$$\psi_n(x', t') = \sum_m c_{nm} u_m(x') \exp(\mathrm{i}Et'/\hbar), \tag{4.41}$$

which is Eq. (27)–[6]. Rearranging the Dirac equation into the form of a Schrödinger equation $H\psi = \mathrm{i}\hbar(\partial/\partial t)\psi$, Heisenberg solved to first order in perturbation theory for the c_{nm} (Eq. (30)–[6]), which he substituted back into Eq. (4.41). To this order of accuracy he calculated $\langle x'k'|R_S|x''k''\rangle$, where he set $t' = t''$ (see Eqs (32) and (33) in [6]), thereby losing Lorentz invariance in the calculation. Heisenberg next isolated divergent terms (light cone singularities) which he related to singular terms expected from S: these terms cancel 'when one goes over from R_S to the matrix r'. The result was a corrected version of Dirac's result from Solvay, which is the 'additional density'

$$\delta\rho = -\frac{1}{15\pi}\frac{e^2}{\hbar c}\left(\frac{\hbar}{mc}\right)^2 \Delta\rho_0, \tag{4.42}$$

where ρ_0 is the external charge or inducing charge density. Heisenberg mistakenly interpreted Eq. (4.42) as having 'no physical significance'.[29]

Actually Eq. (4.42) yields the measurable Uehling effect. Edwin Uehling (1935) chose to explore Eq. (4.42) with Heisenberg's density-matrix procedures rather than with the formulation of Oppenheimer and Furry (1934) because 'it is more convenient to work with expressions which are unconditionally convergent'. He interpreted Eq. (4.42) to be a polarization of the vacuum caused by pair production induced by the external static potential. Consequently deviations from Coulomb's law can be expected. For example, Uehling predicted a displacement of the 2S atomic level in the hydrogen atom owing to a deviation of the potential from Coulomb's law of the amount

$$V = e\int \frac{\delta\rho(r)\,\mathrm{d}r}{r}, \tag{4.43}$$

which is

$$\delta E_{2S} \sim -\frac{\alpha^3}{15\pi} R_y, \tag{4.44}$$

where R_y is Rydberg's constant and the corresponding frequency displacement between the $2S_{1/2}$ and $2P_{1/2}$ energy levels is $\delta\nu = -27Mc$.[30] In Dirac's theory these levels are degenerate. Uehling's result constitutes an argument that vacuum polarization actually consists of a background of virtual electron–positron pairs.

Although in [6] Heisenberg calculated vacuum polarization with the 'intuitive' theory, further considerations of the 'matter density induced by a light wave' and of light–light scattering (which occurs via electron–positron virtual states) require going over to the '*Quantentheorie der Wellenfelder*'. This involves introducing second quantization into the density-matrix formalism. In the course of this analysis Heisenberg (see Eq. (53)–[6]) introduced the basis for charge conjugation, but did not pursue the matter further.

A key point for future considerations is that whereas in the 'intuitive theory of matter waves' the density matrix is written as Eq. (4.39), in the 'quantum theory of wave fields' it is

$$R = \tfrac{1}{2}[\psi, \psi^\dagger] + \tfrac{1}{2}\{\psi, \psi^\dagger\} = \tfrac{1}{2}(R_S + R_F), \tag{4.45}$$

where the ψ's are operators and the physically meaningful (that is finite) matrix is r in Eq. (4.38), which can be written as Eq. (51)–[6] in the zeroth approximation, that is with no external fields. In higher approximations the coefficients c_{nm} in Eq. (4.41) (now interpreted as operators) became functions of the external field strengths like in [2], and S becomes a complicated function of derivatives of the field strengths.

Instead of using time-dependent perturbation theory for calculating r, as he did in the 'intuitive' theory, Heisenberg developed an iteration method for the energy expressed through the density matrix r (Eq. (56)–[6]). Up to first order in e, Heisenberg's method agreed with ordinary perturbation theory (see Wentzel, 1943). Higher-order terms are difficult to calculate because it turns out that

S and not R_S contributes (see Eq. (59)–[6]). He applied this iteration method to order α, where infinities are encountered. For example, Eq. (61) of [6] gives the matrix element for light–light scattering, previously calculated by others in less systematic ways.[31]

At this point, however, a problem developed because Heisenberg calculated that the process where a light quantum creates an electron–positron pair which then annihilates into another photon gives rise to a photon self-energy that diverges logarithmically (see Eq. (69)–[6]). He considered this result to be reasonable by analogy to the self-energy of an electron. Heisenberg ascribed the appearance of the additional infinity to the '"quantization" of the charge distribution', that is to second quantization.[32]

In his concluding paragraph Heisenberg suggested that the '*anschaulich* theory' contains the 'correct correspondence-like description of the process, and the transition to quantum theory cannot be made as primitively as is being attempted in presently available theories' – that is the available second-quantization procedures were insufficient because they introduced yet another infinity, the self-energy of the photon. As a hint to future quantization procedures Heisenberg had in mind the 'Klein–Jordan trick', as will become evident in Heisenberg–Pauli correspondence to be discussed.

Heisenberg concluded that the problem of no adequate second-quantization procedure is reflected in Dirac's theory where there is no 'clean separation of fields'. This manifests itself in the fact that only R_S can be expressed so 'simply by the matter wave functions' (reference here to the charged symmetrized form of R_S in Eq. (4.45)), but not so for the matrix r. In Heisenberg's opinion a unified field theory of light and matter that also fixes α would provide a 'consistent combination of the requirements of quantum theory and those of correspondence with the intuitive field theory [would] be possible'.

4.7.5 *Some reactions to Heisenberg*

On 16 May 1934 Weisskopf wrote to Heisenberg that Pauli's 'disgust was only against the subtraction physics and not against the problem itself' (Pauli, 1985).

On 8 June 1934 Heisenberg sent to Pauli and Weisskopf the manuscript for his 1934 paper for comments. In Heisenberg's opinion although the 'theory is extraordinarily ugly [it] shows that the path from the hole theory to $e^2/\hbar c$ is not too far away' (Pauli, 1985).

Pauli replied on 14 June 1934 that Weisskopf and he were in essential agreement with Heisenberg's work and that the results supported Pauli's 'disgust for the *Limes-Akrobatik* and "*Subtraktionphysik*"' (Pauli, 1985). It seemed that 'all infinities enter first through the field quantization'. For example, Heisenberg's 1934 second-quantized version of Dirac's density-matrix formalism introduced the photon's infinite self-energy.

Pauli continued, 'In physical considerations, I hardly believe that this theory can lead to any results that do not already exist and which can be proven from elementary calculations on the frequency of pair production.' On Heisenberg's statement in [6] concerning 'a modification of the Klein–Nishina scattering formula ... in the region of the Compton wavelength, this will amount to about one-tenth of 1%', that is, as Pauli wrote, 'part per thousand'. But corrections of this sort, continued Pauli, are dubious when predicted by Heisenberg's theory because '*neglect of the radiation reaction force is clearly no longer legitimate in this order of magnitude* and we come to a domain where it is essential to understand the self-energy and $e^2/\hbar c$' (emphasis in original). By neglect of the '*radiation reaction force*' Pauli most likely meant neglect of α^2 contributions to Compton scattering because they contain the self-energy of the electron. Heisenberg and Hans Euler (1936) discussed these contributions (Section 4.12).

On 16 June 1934 Heisenberg replied that Euler and Bernhard Kockel were in the process of calculating the scattering of light by light (an α^2 calculation) for which subtraction physics was unnecessary, and that he was giving further thought as to how the radiation reaction force affected corrections to the Klein–Nishina formula (Pauli, 1985).

Taking account of Pauli's and Weisskopf's new work on the quantization of the Klein–Gordon equation (to be discussed in Section 4.8), in a letter to Pauli of 16 June 1934 Heisenberg proposed a path for future research (Pauli, 1985):

> We pay heed not at all to experiment and correspondence, but ask: how can we generally develop a quantized wave theory which satisfies the following postulates: 1. The number of particles, which correspond to the waves, should not necessarily be constant; 2. Relativistic invariance (eventually gauge invariance); 3. A single particle should be a trivial solution of the fundamental equations (without any infinite self-energy); 4. The energies are positive.

The next step, continued Heisenberg, was a 'unified field theory' of radiation and matter.

Pauli replied on 16 July 1934. Weisskopf and Pauli expected that in their theory as well the self-energy of the photon should be infinite but had not yet accomplished this result (Pauli, 1985). Concerning Heisenberg's 'favorite demand of the month', Pauli disagreed with a charged particle being a trivial solution because of the infinite range of its Coulomb forces. Regarding Dirac's subtraction physics in general, Pauli disagreed with its basic formulation because it yielded a finite vacuum polarization but an infinite self-energy.

> [It was just this striving] that led Dirac to the deplorable 'subtraction physics' which is based on a distinction between results which one obtains with and without quantization of waves. It seems to me that in the derivation of the infinite self-energy of the electron quantization of waves plays only an accidental role . . . On the contrary, there seems to me to stand a far-reaching analogy between charge and energy problems. Instead of the polarization of the vacuum one could also say 'self-charge'. We should demand as substitution for your [point 3 in Heisenberg's letter to Pauli of 16 June 1934] that the expectation value of charge- and energy-densities remain finite and that no distinction is introduced between material and electromagnetic energy.

Pauli further criticized Heisenberg's positron theory; in particular he wrote that Heisenberg's result of a finite vacuum polarization but infinite self-energy for the electron meant that Heisenberg made mistakes in the subtraction process (25 July 1934, in Pauli, 1985). Pauli suggested some alternate routes but was not optimistic. He wrote that the 'true progress of theory is not at all along this path, but through fixing $e^2/\hbar c$ and to seek a folding together of electromagnetic and matter fields (wherein the correspondence-limit *Ansatz*: "we replace p_ν with $p_\nu - (e/c)\varphi_\nu$" would have to be systematically modified)'.

Pauli's letter of 13 October 1934 leveled devastating criticism at Heisenberg's subtraction physics (Pauli, 1985): For example, since the vacuum polarization and the electron's self-energy are analogous, then a theory in which one is finite but the other infinite is 'inconsistent. (The inconsistency comes about psychologically through too strong imprisonment in Dirac's adverse influence in Part I of the work.)' Pauli suggested either abandoning the subtraction physics as 'ugly', or using it to remove the 'polarization of the vacuum *and* the self-energy. Between these possibilities in the Part II of your work you fall into a little ineptness (under Dirac's influence)' (emphasis in original). The editors of the Pauli correspondence rightly concluded that 'to the great disillusionment of all participants the lengthy discussions on hole theory and quantum electrodynamics led to no concrete results'.

As Heisenberg wrote to Pauli on 28 October 1934 (Pauli, 1985), 'With the results of our self-energy discussions I cannot commence to make any reason. The entire subtraction physics contains nonsense and should be replaced by something better. How is your work going on the Bose-electron?' (Reference here to the Pauli–Weisskopf theory of the Klein–Gordon equation which produces charged Bose particles.) Heisenberg emphasized that a theory of electrons and positrons was important for development of a proper nuclear theory: 'I find it a scandal that we are presently too stupid to peel away the true simple formalism behind our complicated hole-mathematics'.

Besides the ill-defined limiting processes for determining matrix elements on the light cone, Pauli had not been comfortable with second quantization. For example, we recall that in a letter to Heisenberg of 18 January 1933 Pauli wrote of his dislike for the manner in which the 'exclusion principle entered with anticommutators. In a future relativistic theory of the many-body problem [second quantization] could be of use. But how, I presently know not'. Hans Jensen (1963) recalled that Enrico Fermi's 1933 theory of β-decay convinced Pauli that 'there was tangible physics in second quantization'. Here it is reasonable to conjecture that Pauli's 'disgust' for Dirac's theory with its infinities led him to grasp at 'application of our field quantization', that is to return to second-quantization methods from the Heisenberg–Pauli quantum electrodynamics in order to seek a new method for exploring the interaction between light and matter.

And this Pauli did in work that would preoccupy him over the next six years. On 14 June 1934 Pauli wrote to Heisenberg about research on a 'curious sort of thing . . . instead of Dirac's, the old scalar Klein–Gordon equation' (Pauli, 1985).

4.8 Quantization of the Klein–Gordon equation: the Pauli–Weisskopf theory

On 14 June 1934 Pauli wrote about the following contrast between the Klein–Gordon and Dirac equations: For the Klein–Gordon equation

$$\left(\Box - \frac{m^2c^2}{\hbar^2}\right)\psi = 0, \tag{4.46}$$

the particle density

$$\rho = \psi^* \frac{\hbar}{\mathrm{i}}\frac{\partial \psi}{\partial t} - \frac{\hbar}{\mathrm{i}}\frac{\partial \psi^*}{\partial t}\psi \tag{4.47}$$

is not positive definite, whereas the energy density

$$H = \hbar^2 \left|\frac{\partial \psi}{\partial t}\right|^2 + \hbar^2 c^2 \,|\boldsymbol{\nabla}\psi|^2 + m^2c^4\,|\psi|^2 \tag{4.48}$$

is positive definite. However, for the Dirac equation

$$\left(\partial\!\!\!/ + \frac{mc}{\hbar}\right)\psi = 0, \tag{4.49}$$

the particle density

$$\rho = \psi^*\psi \tag{4.50}$$

is positive definite, whereas the energy density

$$H = \psi^*[-\mathrm{i}\hbar c\boldsymbol{\alpha}\cdot\boldsymbol{\nabla} + mc^2\beta]\psi \tag{4.51}$$

is not.[33]

Pauli continued in the 14 June 1934 letter to Heisenberg:

> the application of our old formalism of field quantization to this [new] theory leads *without any further hypothesis** (without 'hole' idea, without limiting acrobatics, without subtraction physics!) to the existence of positrons and to processes of pair production with easily calculable frequency. [Emphasis in original.]
>
> *There is (after field quantization) automatically only positive energy! Everything is gauge invariant and relativistically invariant!

Dirac's assertion that a second-order time derivative in a wave equation contradicted the quantum mechanical transformation theory 'is pure rubbish and is refuted by application of our field quantization'.

Unfortunately, continued Pauli, the scalar field theory had 'little to do with reality since spin does not enter in a relativistically invariant manner. [So] practically speaking we cannot proceed much further with this curiosity but it has made me happy that I can again cast aspersion on my old enemy – the Dirac theory of the spinning electron'.[34]

Pauli continued, 'I have been musing over the great question, what is $e^2/\hbar c$?' We recall that, fundamental to the thinking of Heisenberg and Pauli, was that a theory which can fix the value of $e^2/\hbar c$ most likely will have no divergent quantities. Consequently, wrote Pauli, the 'charge concept' is fundamental and so a future theory should connect the gauge group with positive definite energy and the exclusion principle:

> I hope that a strong liberation from Dirac's *Ansätzen und Gedankengängen* is possible. – I wonder whether we cannot introduce spin and the exclusion principle together instead of one after the other. (Lately I believe: first '*anschauliche*' (??!) ψ-waves in ordinary space,† then 'second' quantize them!) I meditate often on these problems . . .. [But] I cannot do better on this point [than with the currently accepted second-quantization methods].
>
> †I believe still, that classical light waves are *anschaulich*, but not ψ-waves, not even if they propagate in ordinary space.

Heisenberg replied on 16 June 1934 to Pauli's and Weisskopf's new results (Pauli, 1985): 'Your theory with the Klein–Gordon equation has me very interested. Perhaps through it we can find the correct method of field quantization'. Heisenberg put his finger squarely on the problem with the c-number single-particle theory of the positron, with the Heisenberg no-holds-barred approach to physics problems:

> I am completely in agreement with you that [owing to second quantization there are] not the least grounds to set $\psi^*\psi$ for the density . . .. Therefore, we will now seek modifications which force the introduction of spinors and Fermi statistics when the Klein–Gordon equation is quantized.

In the subsequently published Pauli–Weisskopf paper [7] (completed July 1934) the point concerning Dirac's single-particle theory was developed further.[35]

Owing to pair production, wrote Pauli and Weisskopf, one can no longer limit oneself to the single-body problem. For example, in view of the mixed terms in the expression for the current density in Eq. (4.9) transposed to occupation number space, 'no reason exists any longer for the special form $\psi^*\psi$ for the charge density'. They went on to interpret the second-quantized particle density from the Klein–Gordon equation as a charge density that can have positive and negative integer eigenvalues.

4.9 Toward a connection between spin and statistics

On 28 June 1934 Pauli thanked Heisenberg for his 'unexpected interest' in the 'old Klein–Gordon equation', and went on to write that in regard to your suggestion of forcing spin into the quantization procedure, 'the result is interestingly negative' because gauge and relativistic invariance are lost, as well as a positive definite energy (Pauli, 1985). This calculation cast some doubt on the generality of

> our formalism for field quantization can in no way be applied generally (i.e., with an arbitrary form of the Lagrangian and Hamiltonian), in agreement with the exclusion principle (with $[\ldots]_+$) as in accord with Einstein–Bose statistics (with $[\ldots]_-$), and that in particular the former [the anticommutator] must be tried out. From a physical point of view it is gratifying to me that no exclusion principle is attainable *without spin* (at least not invariantly and only with positive energy), since the two are so closely connected. I just don't seem to be able to find a reasonable modification of the quantization prescriptions of field theory. What do you think about all this? [Emphasis in original.]

Heisenberg (letter of 11 July 1934) agreed with Pauli's calculations, and added that it was 'logical that without spin we must operate with Bose statistics' (Pauli, 1985). But he had only 'vague conjectures' on Pauli's questions concerning a connection between spin and statistics, and more of his own conjectures on a unified field theory of light and matter. This theory would avoid what Heisenberg referred to as the 'principal joke of the Dirac theory of electrons, which is the demand that the density be $\psi^*\psi$ (transformation theory?!), that forced the existence of spin'.

Amongst Pauli's reply of 16 July 1934 was the following far-reaching conjecture (Pauli, 1985):

> I conjecture a very close connection between spin and the exclusion principle, all the more so because the *far-reaching formal similarity that exists between the*

> *exchange relations of the spin matrices on the one hand and those of the Jordan–Wigner formalism for the quantization of wave amplitudes according to the exclusion principle on the other hand. I have not yet succeeded in utilizing this formal analogy physically*. [Emphasis in original.]

So it was that Pauli's penchant for mathematics provided an important clue toward the connection between second quantization, spin and statistics: The similarity between the 'exchange relations for the spin matrices' ($\{\alpha_i, \alpha_j\} = 2\delta_{ij}$) and 'those of the Jordan–Wigner formalism' for annihilation and creation operators for spin-$\frac{1}{2}$ particles ($\{a, a^\dagger\} = 1$). On 4 November 1934 Pauli sent Heisenberg a copy of the Pauli–Weisskopf paper 'On the quantization of the relativistic wave equation', the paper to which Pauli referred as his 'anti-Dirac paper'. In the accompanying letter Pauli stated what to him was the key point (Pauli, 1985):

> An important and interesting point is the fact that our theory can be implemented only with Bose–Einstein statistics, because here a necessary connection begins to dawn between spin and statistics.

Pauli and Weisskopf brought into play relativity considerations. They proved that if the field amplitudes in the Klein–Gordon equation are quantized with anticommutators, then the four-current does not transform like a four-vector. With all the nice properties of this theory, they wrote, 'One might perhaps be surprised why "nature" has not availed itself of the possibility of oppositely charged particles without spin and with Bose statistics'.

4.10 The connection between spin and statistics

In (1940) Pauli published his next 'anti-Dirac paper', 'The connection between spin and statistics'. He realized how to use the special theory of relativity to explore in a more general manner than previously the space-time transformation properties of wave functions. From the Lorentz transformation properties of wave functions for arbitrary spin, he proved that theories for half-integral spin particles have no positive definite energy, and theories for integral spin have no positive definite particle density. Pauli emphasized that these results follow without assuming that the 'wave equation is of the first order [and consequently] not conclusive [is] the original argument of Dirac'. In fact, Pauli continued, these results demonstrate the inadequacy of a single particle or c-number theory and the necessity for second quantization.

He then determined how the Lorentz transformation properties of wave functions determine the way that bilinear combinations of wave functions are related to the causal Lorentz invariant D-function. Pauli found that it is not possible to include causality in a theory for integral spin that is quantized with anticommutators (he invoked causality in order to exclude the Lorentz invariant D_1-function that does not vanish outside the light cone).[36] The price to be paid for quantizing a theory with half-integral spin according to Bose statistics is a nonpositive definite energy, which is unphysical. Pauli concluded that the 'connection between spin and statistics is one of the most important applications of the special relativity theory'.

The special theory of relativity turned out to be the element necessary to complete the circle connecting intrinsic symmetry, spin and statistics. This result could only have pleased him greatly. We recall Pauli's lifelong interest in relativity theory (a subject in which as a young man he first came to everyone's notice) and his desire for precise mathematical formulations (at which he was thwarted repeatedly in the 'subtraction physics'). Gregor Wentzel (1943), among others, has described Pauli's result as 'one of the most beautiful successes of the quantum theory of fields'.

4.11 Return to 1934

Having brought the union between spin and statistics to a happy ending, let us pick up the strands of development of positron theory.

On 26 November 1934 Heisenberg proposed to Pauli (1985) a scheme to formulate a unified theory of radiation and matter from the density matrix: Until that time the singularities in the density matrix had been determined through the fields. Why not invert the process and determine the potentials and field strengths from the density matrix by limit processes?

The basic equation of Heisenberg's new proposal is

$$A^{\rho}(r') = \frac{\hbar c}{e}\frac{3\pi^2}{2}\sum_{kk'}\alpha^{\rho}_{k'k''}\int \mathrm{d}r''|r' - r''| \cdot (r'k'|R_S|r''k'')\delta(r' - r''), \tag{4.52}$$

which Heisenberg obtained from his (1934a) paper. The function $\delta(r' - r'')$ is not the usual Dirac δ-function; rather it serves to smear out singularities and becomes the Dirac δ-function in the limiting case of infinitely thin width. Other relevant equations are the commutator of R_S's and an equation giving the fine structure constant as a function of volume integrals over the commutator of R_S and of δ-functions.

Pauli replied (30 November 1934) that he had 'brooded' over Heisenberg's letter for twenty-four hours (Pauli, 1985). Pauli's basic disagreement was with

Heisenberg's use of the singular part of the density matrix R_S which has no direct physical meaning (besides not vanishing for the vacuum) instead of the nonsingular portion r which was 'connected directly with the charge density'. Consequently the resulting exchange relations for r are easier to deal with compared to the ones for R_S. This scheme also avoids the complex limiting procedures proposed by Heisenberg for calculating the fine structure constant: 'through permutation of factors (à la Klein–Jordan–Heisenberg) we could try to eliminate the self-energy'.

Involved with calculations on his new program, Heisenberg did not reply to Pauli's letters until 12 December 1934 (Pauli, 1985). Heisenberg defended his belief that in the 'future theory' the matrix R_S should appear and not r. The reason is that in a theory based on r the electromagnetic field quantities would have to be introduced separately because they are not uniquely determined by the charge density: in the r-formalism 'density and field are clearly independent degrees of freedom'. In the R_S-formalism the electromagnetic- and matter-fields can be considered to be 'different properties of one and the same field'. From Eq. (33)–[6] Heisenberg calculated

$$\frac{e^2}{\hbar c} = \lim_{\varepsilon\to 0} \frac{3\pi}{2} \int_0^\infty \mathrm{d}p p^3 f^2(\varepsilon, p); \tag{4.53}$$

instead of the δ-functions from Eq. (4.52), Heisenberg used a function f, where $\int p\,\mathrm{d}p f(p, \varepsilon) = 1$.

Pauli replied (9 February 1935) that he did not believe that the function f could ever be made unique (Pauli, 1985).

On 22 March 1935 Heisenberg suggested replacing the radiation fields in the Hamiltonian with their expressions in the density matrix (Pauli, 1985). This led to a Hamiltonian quadratic in the density matrix which was considered 'ugly' and 'horribly complicated'. Since it is quadratic then two solutions emerged that can be interpreted as the electron and light quantum.

Heisenberg wrote to Pauli on 25 April 1935 that the situation has become similar to the one in 1922, '*Wir wissen dass alles falsch ist*'. In the *Anlage* to this letter Heisenberg wrote that with regard 'to the discussion of these difficulties [particularly the electron's self-energy] we should remember Bohr and Rosenfeld's analysis from which emerged the result that simple indeterminancy relations exist only for the average values of field strengths over a definite space-time region'. Heisenberg proposed a formulation of the quantized density-matrix theory where all δ-functions that gave rise to light cone singularities are replaced with a nonsingular function Δ that has a finite width, thereby smearing out singularities. In this new quantum electrodynamics analysis of measurement interferences will be as important as in quantum mechanics. To

Pauli's inquiry whether Heisenberg's theory can produce the hydrogen atom spectrum, Heisenberg replied that he believed so.

On 15 June 1935, at the end of a letter in which Pauli reported to Heisenberg the paper of Einstein with Boris Podolsky and Nathan Rosen '(on the whole not good company [for Einstein])', Pauli wrote that Heisenberg's idea 'to introduce Δ as an influence of the measurement apparatus' was unsatisfactory (Pauli, 1985). By mid-November 1935 Heisenberg had abandoned this approach.

4.12 Light–light scattering

The calculations which Euler and Kockel began in summer 1934 (Section 4.7.5) were arduous and were not published until February 1935 (Euler and Kockel, 1935). The possibility of light–light scattering via pair formation cannot be described by a linear classical electromagnetic theory wherein the superposition principle is valid. Euler and Kockel showed that to order α^2 this process could be calculated by replacing the Hamiltonian for matter and field by one that contains only the radiation field (a phenomenological Hamiltonian):

$$H = \frac{1}{8\pi}\int(\boldsymbol{B}^2 + \boldsymbol{D}^2)\,\mathrm{d}V - \frac{1}{360\pi^2}\frac{\alpha}{E_0^2}\int\{(\boldsymbol{B}^2 - \boldsymbol{D}^2)^2 + 7(\boldsymbol{B}\cdot\boldsymbol{D})^2\}\,\mathrm{d}V, \tag{4.54}$$

where $\boldsymbol{D}$ is the dielectric displacement, $\boldsymbol{B}$ is the magnetic induction, and E_0 is the field at the 'edge of the electron' where quantum electrodynamics is assumed to break down ($E_0 = m^2c^3/e\hbar$). Eq. (4.54) shows the nonlinearity of the process. The additional term in { } they 'intuitively' interpreted as the interaction energy of the light quanta.

In December 1935 Heisenberg and Euler completed a detailed investigation of light–light scattering. They emphasized that the very process of pair production, whether real or virtual, changes the Maxwell equations. Using the 'intuitive wave theory' from [6] for the Hamiltonian expressed through the r-matrix, they carried out an incredibly detailed calculation for the energy density and the Lagrange function for parallel constant electric and magnetic fields imposed on the vacuum. Their result for the Lagrange density is

$$L = \frac{1}{8\pi}(\boldsymbol{D}^2 - \boldsymbol{B}^2) + f(a, b), \tag{4.55}$$

where $f(a, b)$ is a function of dimensionless numbers a and b that occur in the form $a^2 - b^2$ and $(ab)^2$, like the two Lorentz invariants $\boldsymbol{D}^2 - \boldsymbol{B}^2$ and $(\boldsymbol{D}\cdot\boldsymbol{B})^2$, and represent the ratio of the external electric and magnetic fields, respectively, to the fields at the edge of the electron. Euler and Kockel's result emerged for $a \ll 1$, $b \ll 1$.

Heisenberg and Euler continued: 'The results [from the "intuitive wave theory"] cannot be transferred without further ado' to the quantum theory of wave fields because virtual pair production must be discussed. With reference to the perturbation method that Heisenberg introduced in [6] for the Hamiltonian expressed through the density matrix, they discussed qualitatively calculations of the Compton effect and light–light scattering until divergences are encountered. For example, the effective cross-section for Compton scattering calculated to order α converges, whereas the one calculated to α^2 diverges, owing to the self-energies of the photon and the electron. But the success of the Klein–Nishina formula indicated that neglect of second- and higher-order contributions in α leads to correct results. Consequently the 'intuitive' theory can be taken directly into the quantum theory of waves, including the nonlinear effective energy density, which can be interpreted as the result of scattering processes (order α^2 light–light scattering converges) with finite effective cross-section (order α^3 light–light scattering diverges).

Are the results on light–light scattering definitive? No, they wrote, because the theory of the positron and the current quantum electrodynamics were undoubtedly to be regarded as provisional. For example, the methods for calculating the subtraction matrix S were still somewhat arbitrary.[37,38]

4.13 Weisskopf and vacuum polarization: a new statement about infinities

In an erudite paper of 1936, Weisskopf [8] presented a critical overview of positron theory and deduced the Euler–Kockel results in a more straightforward manner without the density matrix. On this calculation Heisenberg wrote to Pauli (18 February 1936): 'that would provide beautiful support for the unequivocality of the subtraction terms'.

Weisskopf postulated that 'Heisenberg's subtraction methods' are identical to assuming 'three properties of vacuum electrons as being physically meaningless' (see Section I of [8]) because 'all three quantities prove to be divergent sums after summing the contributions from all vacuum electrons'. This being the case then Heisenberg's 'subtraction methods are significantly less arbitrary than has hitherto been assumed in the literature'.

To substantiate this point Weisskopf used time-dependent second-order perturbation theory to calculate vacuum polarization (see Eqs. (32)–(34) in [8]).

Invoking assumptions I_2 and I_3 (see [8]), divergent terms were discarded and the finite contribution in Eq. (34)–[8] resulted. He carefully noted that perturbation theory permits one only to recognize what terms are divergent, whereas Heisenberg's density-matrix method enables calculation of the form of the divergent terms and exhibits 'relativistic invariance and the validity of the conservation laws'.

Especially prescient is Weisskopf's comment: 'It may also be noted that the polarizability could in no way be observed, but would multiply all charges and field strengths by a constant factor'.[39] This would be a key result of a renormalized quantum electrodynamics in which subtraction is replaced by extraction of an infinite constant that multiplies the free electron's bare charge e to give the measured charge e_1:

$$e_1 = Z_3^{1/2} e, \tag{4.56}$$

where Z_3 is the infinite constant from vacuum polarization.[40]

Earlier in this paper Weisskopf argued that explicit inclusion of interactions between electrons would require one 'to quantize the wave fields. As is well-known, this already leads to divergences even without assuming infinitely many vacuum electrons and will not be touched upon in more detail in the following discussion'. Weisskopf's message was to avoid second quantization. He concluded with a proposal for continued exploration with the 'hole theory of the positron [which] has brought with it no essential difficulties for electron theory as long as one limits oneself to treating unquantized wave fields'.

4.14 Some other approaches to quantum electrodynamics: cosmic ray physics, nonlinear theories and the lattice world (*Gitterwelt*)

Having abandoned the program of combining measurement theory with quantum electrodynamics, in a letter to Pauli of 23 May 1936, Heisenberg presented another approach to quantum electrodynamics based on the assumption that heavy and light particles could be considered separately (Pauli, 1985). In a theory where all masses are zero, wrote Heisenberg, there should be only one natural constant with a dimension, the velocity of light, c. Taking the Pauli–Weisskopf theory as a guide, the 'simplest model' in Heisenberg's view was a nonlinear wave equation

$$\Box\psi = \lambda\psi\psi^*\psi, \tag{4.57}$$

where λ is a pure number that determines the types of different solutions. Is there a value of λ for which the solutions are simple?:

> Let us assume that there exists a theory for electrons + radiation, without protons, neutrons, neutrinos, etc. etc, and further that $e^2/\hbar c = \lambda$ has been determined in the manner sketched above. Then the addition of new particles, which interact with the old ones, would destroy, everything, because they would again change the solution type. Therefore, for $m \to 0$, this theory must contain the entire particle salad (electrons, neutrinos, neutrons, protons, light quanta). We can bring this argument into the following simple form: The theory of elementary particles is *not* linear; we cannot superimpose solutions, and so it is not a theory in which electrons and neutrons can finally be grouped together. Thus, in the definitive theory $e^2/\hbar c$ and Fermi's constant g must be determined all at once . . . . Nature probably overcomes the self-energy difficulty in the following refined manner: for a theory: electron + radiation the self-energy [Coulomb] diverges as $\int \mathrm{d}k$; if we introduce the positron the divergence is only $\int \mathrm{d}k/k$, if we include also neutrinos, too, convergence as $\int \mathrm{d}k/k^2$ (???). But naturally this is phantasy.

Not quite, because Heisenberg went on to write that the

> problem of elementary particles is a mathematical one, namely, simply the problem of how we can construct a nonlinear, relativistically invariant and quantized wave equation without any natural constants. It seems to me not implausible from the above argument on the self-energy that we require already a considerably abundant particle assortment in order to satisfy all requirements of such a theory.

Beginning in 1936 Heisenberg began extending the Fermi β-decay theory into cosmic ray phenomena (Heisenberg, 1936). Second-order perturbation theory applied to the high energy collision of a proton with a nucleus yields cross-sections that increase with energy, and describe high energy processes as an 'explosion' within a small volume in which many final state particles are produced.[41,42]

Since large energies correspond to interactions over small distances, in (1936) Heisenberg was led to introduce a 'universal length that must perhaps be connected with a new change of principle of the formalism, similarly as, for example, the introduction of the constant c has given a modification of the prerelativistic formalism'. The length is of order of magnitude $f^{1/2} \sim 10^{-17}$ m (see note 42). For the purpose of providing a qualitative discussion of the explosion, Heisenberg used a semiclassical version of the Fermi interaction of the form $g\psi\psi^*\psi$, from which emerged his suggestion to Pauli of a nonlinear field equation (letter of 23 May 1936).

Pauli responded positively to this suggestion (letter to Heisenberg, 26 October 1936, Pauli, 1985). Pauli discussed results of their work on a version of Heisenberg's earlier '*Gitterwelt*'. He had been engaged in this research since about June 1936, and by October he was losing interest. Pauli suggested that the situation in quantum electrodynamics was similar to the one in the quantum mechanics of the early 1920s when perturbation theory failed to produce the stationary states of the helium atom. This problem turned out to be deeper than perturbation theory. But in those days everyone searched first for new mechanical models, whereas today everyone searches for new Hamiltonian operators. It would clarify matters if after 'once again returning from Helgoland' you would declare that 'for a system with infinitely many degrees of freedom the eigenvalues of the Hamiltonian operators are "fundamentally not observable" (in the same sense as the classical rotation frequencies in the "orbits" in the old quantum theory)'.[43]

Heisenberg (see letter to Pauli of 29 October 1936 in Pauli, 1985) agreed that the situation was like the one in the 'year 1923 with helium [but now] the entire quantum theory of wave fields is wrong'. Heisenberg agreed that they were working with too many observables and, to make things worse, with a half-classical theory to deal with cosmic ray showers. Nevertheless, before 'returning' to Helgoland he presented to Pauli for comments yet more lattice world models with their fictitious Hamiltonians.

Heisenberg described their collaboration to Born (letter of Heisenberg to Born, 3 November 1936, quoted from Cassidy, 1981):

> Pauli continually tries to prove that wave quantization always diverges, even with a fundamental length; I also silently believe that Pauli is right, but temporarily maintain the opposite, and in this way we get to know the mathematical properties of a nonlinear quantum field theory which are highly interesting.

By November 1936 Pauli had convinced himself that inclusion of a fundamental length could not prevent infinite cross-sections in any model of shower production (see letter of Pauli to Heisenberg, 20 November 1936, in Pauli, 1985). By spring 1937 new improved cosmic ray data separated into hard and soft components revealed that every shower is a cascade. For the next decade Heisenberg maintained the hope that a few shower events would turn out to be explosions (Pais, 1986). The reason could only have been that explosive events would lend credence to a fundamental length.

On 2 May 1937 Pauli resigned from the fundamental length program which by then also included the building of lattice world models. The fundamental length and the *Gitterwelt* did not lead to finite energy values and the mathematics had become intractable; for example, quantizing *Gitterwelt* Hamiltonians (even a simple version with a Dirac equation). Moreover, the relativistic invariance of such a theory was not possible owing to the universal length (see letter of Pauli to Heisenberg, 2 May 1937 in Pauli, 1985).

4.15 The universal length

Heisenberg, however, persisted because, in his opinion, the universal length offered a means for discovering 'new conceptual structures' [10]. Besides providing a concise summary of the state of cosmic ray physics in 1938, Heisenberg's paper 'The universal length appearing in the theory of elementary particles' illustrated his attempt to use in 1938 the successful methodology of 1925, tried again in 1932 in nuclear physics (see Section 5) – moving across new frontiers required demarcating off their boundaries with correspondence-like limit procedures. In 1938 the boundary was set by the universal length which Heisenberg took to be of the order of the electron's radius.

Section I of [10] includes an excellent analysis of pitfalls in correspondence-limit arguments for passing between classical, relativistic and quantum physics. Planck's constant and the velocity of light are considered as fundamental constants of the 'first kind' because they designate the 'limits which are set to the application of intuitive concepts'. So, moving across the frontier that separated Newtonian physics and the Maxwell–Lorentz theory from relativistic theory required realization that c is a fundamental constant. Similarly for statistical physics where a quantum theory required realization that h is a fundamental constant. Heisenberg referred to these transitions denoted by characteristic constants of nature as 'conjoining' theories without contradictions.

Then conjoining quantum mechanics and relativistic wave theory required a new universal constant of the first kind with length dimension because in relativistic quantum mechanics the fundamental problem is divergences which occur at high energies or small distances (Part III). Evidence for a fundamental length, claimed Heisenberg, had appeared early in applications of quantum electrodynamics in processes other than cosmic ray showers, for example divergent integrals in self-energy calculations were cut off at a 'length of the order of r_0 (or at the corresponding momentum [$137mc$])'. New 'nonintuitive features' in addition to those from quantum mechanics, 'are introduced into physics through the constant r_0'.[44] The new physical phenomena which were expected to appear at distances below r_0 'would lead to convergence of these integrals'. Heisenberg pointed out that even if only a few cosmic ray showers were explosions, the need for a new fundamental length 'will also force the formation of completely new concepts which will find an analog neither in quantum theory nor in relativity theory' (Section V of [10]).

As for the problem of the relativistic invariance of a theory with a fundamental length, Heisenberg wrote that 'it does not seem to matter whether one speaks of a universal length or mass'. For purposes of relativistic invariance one would want a universal rest mass (see Parts II and IV). The mass spectrum, however, could not be derived from Planck's constant and the velocity of light; a length is required. However, continued Heisenberg, a universal length has greater physical significance than a universal mass because it 'signifies a limit to the application of intuitive concepts'. In order to substantiate this point Heisenberg presented the following argument: Since a finite electron mass cannot be deduced from quantum electrodynamics, the definitive theory 'must contain not only $\hbar$ and c but also a universal length r_0, that is, elements which have nothing to do with electrodynamics and quantum theory'. Here, using a dimensional analysis, Heisenberg argued that the question of finite electron mass basically had nothing to do with the electron's charge because one can write mass as

$$m_0 \sim \frac{\hbar}{c r_0}. \tag{4.58}$$

In turn, this means that r_0 will introduce 'new features' that initially 'have no connection at all with the question of the electronic charge', which means that 'new features of the description of nature' due to r_0 will have to be clarified before even a theory for the fine structure constant could be found.

4.16 The infrared catastrophe

Whereas the self-energy divergence is an ultraviolet catastrophe that occurs also in classical electron theory, the divergent result for the emission of small frequency light quanta by electrons is the nonclassical infrared catastrophe. A perturbation expansion in powers of α for scattering of an electron by the Coulomb field of a nucleus leads to a logarithmically divergent cross-section when the scattering is accompanied by emission of a low energy (soft) light quantum. On 10 June 1937 Pauli wrote to Heisenberg about having received a manuscript from Felix Bloch and Arnold Nordsieck (1937) in which the ultraviolet problem, on which Heisenberg and Pauli had done some work, was resolved. Heisenberg wrote back to Pauli (12 June 1937): 'The solution of the problem is surprisingly simple. I am ashamed of myself that I had not solved it'.

Bloch and Nordsieck's method for summing contributions from multiple emissions of soft photons is analogous to the classical physics expansion of the Lorentz force in powers of $r_0/\lambda = e^2 v/mc^3$ (see Eq. (3.34)). They consider only the nonrelativistic case where $e^2 v/mc^3 \ll 1$. For high frequencies the Bloch–Nordsieck method gives the same divergent results as a perturbation series expansion in α.

Pauli was not completely satisfied with the care that Bloch and Nordsieck had taken with their solution, particularly with energy conservation (see note 13 in Bloch and Nordsieck, 1937) and, with Marcus Fierz, he set out to rework the problem more rigorously (see §2 of [9] for a review of Bloch and Nordsieck's results). Their methods demonstrate a new direction in quantum electrodynamics.

In a letter to Weisskopf (20 July 1937), Pauli described results thus far (Pauli, 1985): 'Besides the known difficulties of quantum electrodynamics which concern the self-energy of a charged particle, there exists a divergent result that concerns the emission of light quanta of very small frequency'. This process always accompanies the deflection of an electron by a force field in which the cross-section dq for a particle to undergo an energy loss that lies between E and $E + dE$ is $dq = \text{constant} \times dE/E$.[45]

Always connecting the woes of quantum electrodynamics to the electron's divergent self-energy, we read in the Pauli–Fierz paper [9] that the result of their work 'depends so strongly on the extension of the charged body, however, that a direct application of the result to real electrons is not possible'. As Pauli put it in a 20 July 1937 letter to Weisskopf (Pauli, 1985), 'As for real electrons, God knows!' Pauli was astonished that whereas the appearance of a cutoff frequency usually meant the insolubility of a problem in quantum electrodynamics, this was not the case for the emission of low-frequency light quanta. In the Pauli–Fierz method the cutoff frequency to prevent logarithmic divergence is $\omega_1 = 2\pi c/r_0$.

Carefully delineating the approximations involved in summing up nonrelativistically the contributions to the scattering cross-section from emission of a large number of low-frequency light quanta led Pauli to classical considerations concerning the mass of the electron.

In turn this led to more exact specification of the masses in the electron's Hamiltonian. To illustrate this point in [9] let us speculate on the equation between Eqs. (12) and (IV) in [9], which contains the expression

$$\frac{p^2}{2m}\left(1 - \frac{\mu}{m}\right), \tag{4.59}$$

where m is the inertial mass of the electron and μ is its electromagnetic mass. Whiggishly we could say that Pauli and Fierz missed a key point when they wrote, 'According to our presupposition (I) it is logical to neglect the second term here. . .'.

Perhaps Pauli and Fierz should have taken Kramers's advice in [11] 'that the quantity that is introduced as the "particle mass" is from the very beginning the experimental mass'. In this case Pauli and Fierz would have written the above equation as (to a first approximation)

$$\frac{p^2}{2m}\left(1-\frac{\mu}{m}\right)=\frac{p^2}{2m_{\mathrm{exp}}}, \tag{4.60}$$

where

$$m_{\mathrm{exp}}=m+\mu. \tag{4.61}$$

Might it not have been the case that the association of m with the electron's bare mass and the electromagnetic mass μ with the additional infinite contribution from the electron's virtual cloud of quanta could have followed? But Pauli and Heisenberg were reluctant about having infinities in a physical theory.

5 Theories of the nuclear force in the 1930s

5.1 Heisenberg invents exchange forces in nuclear physics: the metaphor of forces transmitted by particles

In a series of papers in 1932–3 entitled, 'On the structure of the atomic nucleus. I,' 'II', 'III', Heisenberg sought to frame a theory of the nuclear force based on the newly discovered neutron and which was applicable to the origin of the electrons in β-decay.[46]

Introducing the neutron into the nucleus, wrote Heisenberg in (1932a), leads to an 'extraordinary simplification for the theory of the atomic nucleus'.[47] The existence of the neutron permitted him the means to relate the problem of β-decay to the form of the attractive force that binds the nucleus. As Heisenberg wrote to Bohr on 20 June 1932, 'The basic idea is to shift the blame for all principal difficulties onto the neutron [divergent self-energies too] and to refine quantum mechanics in the nucleus' (Pauli, 1985).

But first Heisenberg had to decide whether the neutron was a composite particle consisting of a proton and an electron or a fundamental particle.[48] In either case, Heisenberg assumed that the neutron had spin $\frac{1}{2}$ in order to provide N^{14} with the correct statistics, that is comprising seven protons and seven neutrons, an even number of fermions. But according to the uncertainty principle, the neutron could not be a proton–electron bound state because the composite neutron's binding energy would have to be of the order of $137mc^2$ (m being the mass of the electron), which is one hundred times greater than the measured neutron–proton mass difference (Heisenberg, 1932b). On the other hand, continued Heisenberg, a 'meaningful definition of the concept of binding energy is impossible for the electron in the neutron on account of the denial of the energy law for β-decay'. In 1930 Bohr suggested that owing to the nuclear electrons' continuous spectrum in energy for a supposedly two-body final state, energy conservation is invalid in β-decay (Bohr, 1932). Consequently the fact that the 'stability of the neutron cannot be described by the accepted theory permits separation of quantum mechanics into accessible and inaccessible domains', where the usual quantum mechanics does not apply to nuclear electrons (Heisenberg, 1933a). Heisenberg went on to emphasize that, whereas either the composite or fundamental neutron was adequate for discussing the properties of lighter nuclei, the composite neutron is required for heavier nuclei which are β-emitters.[49]

However, a nonelementary neutron requires its electron to obey incorrect statistics, and so 'it does not appear suitable to delineate such a picture [*Bild*] in more detail'.[50] If under the proper circumstances a fundamental neutron decays into a proton and electron, then the 'conservation laws of energy and momentum are probably no longer applicable'. At this time Heisenberg had not yet accepted Pauli's neutrino hypothesis, which permits application of these two conservation laws with the result that the proton, electron and neutrino can share the energy and momentum in a variety of ways that are characteristic of a three-body final state.[51]

Heisenberg equivocated on whether the neutron is a fundamental particle owing principally to his desire to combine the problem of where the electrons originated in β-decay with the form of the nuclear force, which is clearly unlike the Coulomb force. And so, when necessary, he invoked arguments based on conservation of energy, for example for discussing the stability of certain nuclei against β-decay. But then, at the paper's conclusion, Heisenberg suggested that for certain processes such as the scattering of light from nuclei, it is useful that the neutron is a composite particle[52] (see Brown and Moyer, 1984).

Playing all ends against the middle in this theoretical free-for-all offered Heisenberg an 'arbitrariness [in seeking] a formulation that will not lead sooner or later to internal difficulties'.[53]

By 1932 all that was known about the nuclear force was that it had to be of extremely short range, less than 10^{-15} m. The property of nuclear saturation wherein nuclear densities are constant was accepted although not yet satisfactorily verified. Saturation can occur through nuclear particles interacting closely only with a few other neighboring nucleons, thereby preventing collapse of the nucleus. Exchange forces can accomplish this goal through limiting or saturating the number of bonds between particles. (Another property of nuclear saturation is that the nuclear radius is proportional to the cube root of the mass number.)

To express saturation mathematically Heisenberg (1932a) drew an 'analogy' between the attractive force of a proton and neutron and the exchange force that is a dominant factor in a molecule's stability. For example, in the solution to the problem of the stability of the H_2^+ ion, a strictly quantum mechanical contribution to the energy of the ion can be described metaphorically as the electron being shared or exchanged between the two protons at a frequency that is equal to the exchange energy divided by Planck's constant. The exchange contribution to the helium atom and the H_2^+ ion arises through the indistinguishability of the electrons and so any depiction of this force 'should not be taken seriously' (Bethe and Salpeter, 1957). The inability of the old Bohr theory to account for the stability of the H_2^+ ion presaged the theory's fundamental problems. The stability of the ion is due to an intrinsic property that is not visualizable because the exchange force cannot be developed from intuitions constructed from the world of perceptions, that is our 'ordinary intuitions'.

The exchange force was another great invention in quantum theory that Heisenberg introduced in order to understand the spectrum of the helium atom (see Eqs. (1.6) and (1.7)). Walter Heitler and Fritz London (1927) used Heisenberg's concept of the 'exchange phenomenon [*Austauschphänomen*]' in their theory of homopolar bonding in molecules. Heisenberg (1928) extended the work of Heitler and London toward 'clarifying ferromagnetism'. In (1928) Heitler wrote that it is as 'yet incomprehensible what exchange in reality means'.

As if in response to Heitler, Heisenberg extended the exchange force into the nucleus in a way that offered visualizability and thus, in this case, an understanding of what the exchange force 'in reality means'. Modern nuclear physics and elementary particle physics begins thus (Heisenberg, 1932a):

> If we bring the neutron and proton to a separation comparable to nuclear dimensions then, in analogy to the H_2^+-ion, a migration [*Platzwechsel*] of negative charge will occur, whose frequency is given by a function $J(r)/h$ of the separation r between the two particles. The quantity $J(r)$ corresponds to the exchange [*Austausch*] or more correctly migration integral [*Platzwechselintegral*] of molecular theory. This migration can again be made more intuitive [*anschaulich*] by the picture [*Bild*] of electrons that have no spin and follow the rules of Bose statistics. But it certainly is more correct to regard the migration integral [*Platzwechselintegral*] $J(r)$ as a fundamental property of the neutron–proton pair, without intending to reduce it to electron motions.

Yet in the 1932–3 nuclear physics papers Heisenberg discusses nuclear electrons that 'follow the rules of Bose statistics'. Heisenberg's switch of terminology from 'exchange' (*Austausch*) to 'migration' (*Platzwechsel*) emphasizes the new concept to follow because he had something else in mind for the neutron–proton force. It is reasonable to conjecture that with this switch from exchange to migration, Heisenberg's visualizability (*Anschaulichkeit*) of the migration (*Platzwechsel*) of the electron in the neutron–proton force is, however unintentionally, the visualizability in Figure 5.1(g), where the quantity $J(r)$ is the attractive force between a fundamental nuclear proton and a composite nuclear neutron. Had Heisenberg, himself, depicted the visualizability for the nuclear exchange force from his vivid depictions of it, it would have been the one in Figure 5.1(g).

The attractive force operates through 'migration' of charge from a neutron to a proton which, capturing the Bose electron, becomes a neutron. For this reason I have rendered *Platzwechsel* as 'migration'. 'Change of place' is inappropriate because in Heisenberg's theory the neutron and proton do not merely change places. The metaphor of motion is of the essence here. Heisenberg had taken the first step toward regaining the visual component of visualizability. 'Migration' also conveys the meaning of *Platzwechsel* that enabled Heisenberg's vision of the nuclear force to be brought to fruition in 1935 by Hideki Yukawa. But we are moving ahead of ourselves.

Heisenberg proposed to characterize each particle in the nucleus with five variables: three position coordinates, a spin variable σ^z and a 'fifth number' ρ^ζ, where $\rho^\zeta = +1$ for a neutron and $\rho^\zeta = -1$ for a proton. The 'fifth number' ρ permitted Heisenberg to describe the nucleus as if it were composed of identical particles of spin $\frac{1}{2}$ that obeyed the Pauli exclusion principle; thus, the neutron and proton are two different states of the nuclear particle (or nucleon). In (1937) Eugene Wigner dubbed this 'fifth number' isotopic spin.

Continued Heisenberg (1932a), 'Because of migration, transition elements from $\rho^\zeta = +1$ to $\rho^\zeta = -1$ also occur in the Hamiltonian function', which is

$$\begin{aligned} H = {} & \frac{1}{2M}\sum_k p_k^2 - \frac{1}{2}\sum_{k>l} J(r_{kl})(\rho_k^\xi\rho_l^\xi + \rho_k^\eta\rho_l^\eta) \\ & - \frac{1}{4}\sum_{k>l} K(r_{kl})(1+\rho_k^\zeta)(1+\rho_l^\zeta) \\ & + \frac{1}{4}\sum_{k>l}\frac{e^2}{r_{kl}}(1-\rho_k^\zeta)(1-\rho_l^\zeta) - \frac{1}{2}D\sum_k(1+\rho_k^\zeta), \end{aligned} \tag{5.1}$$

where M is the proton or neutron mass, $r_{kl} = |r_k - r_l|$ is of the order of nuclear dimensions, or 10^{-15} m, p_k is the momentum of particle k, and

$$\rho^\xi = \begin{pmatrix} 0 & 1 \\ 1 & 0 \end{pmatrix}, \quad \rho^\eta = \begin{pmatrix} 0 & -\mathrm{i} \\ \mathrm{i} & 0 \end{pmatrix}, \quad \rho^\zeta = \begin{pmatrix} 1 & 0 \\ 0 & -1 \end{pmatrix}. \tag{5.2}$$

That is, the ρ's are represented as Pauli spin matrices, the eigenvalues of the matrix ρ^ζ represent a neutron (+1) and a proton (−1), J and K are the exchange forces for neutron–proton pairs and neutron pairs, respectively (Heisenberg assumed that J is larger than K), e^2/r is the Coulomb interaction between proton pairs (he neglected any attractive force between nuclear protons), D is the mass defect between protons and neutrons, and the summations are over all particles in the nucleus.[54]

Heisenberg's exchange force operates through transfer of charge between the neutron and the proton. In 1933 at Solvay he showed that the exchange force term (which exchanges charge) can alternatively be written without the ρ-matrices (that is without isotopic spin) as

$$-J(r_{kl})P_{kl}, \tag{5.3}$$

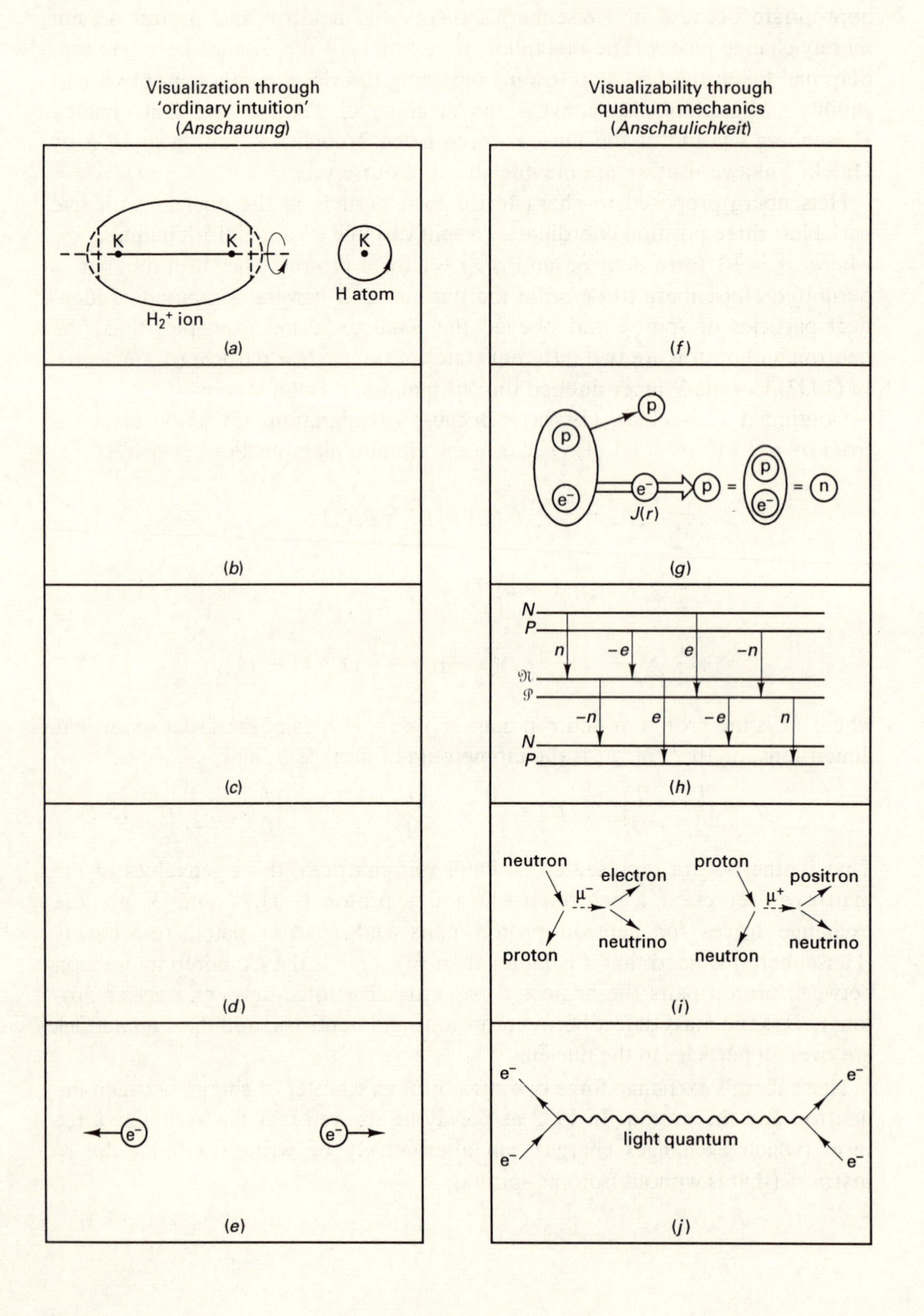
Visualization through
'ordinary intuition'
(Anschauung)
Visualizability through
quantum mechanics
(Anschaulichkeit)
K
K
K
H atom
H2+ ion
(a)
(f)
p
p
e−
e−
J(r)
p
p
e−
n
(b)
(g)
N
P
n
−e
e
−n
𝔑
𝔓
−n
e
−e
n
N
P
(c)
(h)
neutron
electron
μ−
neutrino
proton
proton
positron
μ+
neutrino
neutron
(d)
(i)
e−
e−
e−
e−
e−
e−
light quantum
(e)
(j)

where P_{kl} represents the operator that exchanges the space and spin variables of neutron and proton, and the minus sign results from the extended Pauli principle according to which the total wave function changes sign if all five coordinates of the two particles are interchanged.

For considerations other than *Platzwechsel* and β-decay, the 'neutron can be interpreted as a solid elementary building block of the nucleus'. In the literature the quantities $J(r)$ and $K(r)$ were referred to as 'exchange forces'. Heisenberg's exchange forces are static and central forces of short range. The expressions most frequently used are

$$J(r) = a\mathrm{e}^{-br}, \qquad (5.4)$$

where a and b are adjusted to fit empirical data such as binding energies (for example, see Heisenberg, 1933b, Gamow, 1937 and Brink, 1965).

In summary, in order to describe the *attractive* force between a neutral neutron and a charged proton, Heisenberg extended the metaphorical exchange force from the quantum theory of molecules to an *Anschaulichkeit* (visualizability) of the 'migration' of a Bose electron. In Heisenberg's theory of the nuclear force, the composite and elementary neutron stand side by side.

But there were basic problems with Heisenberg's nuclear force, among them is that in (1933a) he found that his nuclear theory achieved saturation only by introducing short range repulsive forces. In this way the saturated subunit is the α-particle and not the less tightly bound deuteron (see, also, Heisenberg's comments on this problem at Solvay).

Figure 5.1. Two rows of frames show how quantum mechanics distinguishes between visualization and visualizability. Thus, frame (f) is empty, as are frames (b), (c) and (d). (a) is taken from Pauli's (1922) unsuccessful attempt to deduce the stationary states of the H_2^+ ion from Bohr's atomic theory (see Section 1.2). K denotes the central positive charge. In (g) a compound neutron comprising a Bose electron and fundamental proton decays. The Bose electron 'migrates' over to a nuclear proton that captures it and becomes a composite neutron. (h) is from Wentzel's (1936) depiction of β-decay as a 'cascade-like' process analogous to the transition between atomic energy levels. N(P) means a neutron (proton); $\mathfrak{N}$ ($\mathfrak{P}$) means a 'hypothetical stationary state' of a neutron (proton). (i) is from Wentzel's 'schemata' for depicting β-decay as a 'two stage' process that is intermediated by a meson. (e) is the representation of the repulsive interaction between two electrons based on our ordinary intuition (Anschauung). (j) is the new Anschaulichkeit for the Coulomb repulsion between two electrons; two electrons interact through exchange of a light quantum. Today the words 'exchange' and 'migration' are synonymous. Consequently the changing notions of physical reality that began with Heisenberg's remarkable extension of the exchange force from molecular physics to nuclear physics are in frames (g)–(j). Comparison of frames (e) and (j) illustrates the startling contrast between physical reality according to the world of perceptions and the atomic domain (Miller, 1991).

5.2 Ettore Majorana's nuclear theory

The first person to attempt to cure ills of the sort just encountered was the enigmatic Italian physicist Ettore Majorana who had been in correspondence with Heisenberg on the problem of nuclear forces and to whom Heisenberg had sent a prepublication copy of his (1933a) paper. Later in (1933), Majorana set out to formulate a 'new interpretation of Heisenberg's nuclear theory [that avoids] the troublesome ρ-spin coordinate'.[55]

Majorana's guideline was to achieve 'in the simplest possible way the most general and obvious properties of the nucleus', particularly nuclear saturation with the α-particle as the saturated subunit. For qualitative results it sufficed to retain only the proton–neutron force. Majorana sought to avoid adding to the long range attractive force enormously strong short range repulsive forces dependent on the radii of neutrons and protons. Heisenberg had to add such forces in order to achieve saturation at the α-particle. Majorana considered this tack to be 'aesthetically unsatisfactory'. To this end he postulated an exchange force with a sign opposite to that of Heisenberg's and which exchanged the position coordinates of neutron and proton, where Majorana considered only fundamental neutrons. This prescription satisfies Majorana's criterion: 'How can we obtain a density independent of the nuclear mass [nuclear saturation] without obstructing the free movement of the particles by an artificial impenetrability?' Using the Thomas–Fermi method Majorana demonstrated that his formalism led to nuclear saturation.

At Solvay, Heisenberg (1933b) characterized the Majorana exchange force thus: Including isotopic spin,

$$J(r_{kl})\tfrac{1}{4}[(\rho_k^\xi \rho_l^\xi + \rho_k^\eta \rho_l^\eta)][1 + \sigma_k \sigma_l], \tag{5.5}$$

and not including isotopic spin,

$$-J(r_{kl})P_{kl}, \tag{5.6}$$

where P_{kl} is the operator corresponding to a permutation of the spatial coordinates r_k and r_l.

The exchange of charge in Heisenberg's exchange force limits the number of bonds to one between neutrons and protons, thereby naturally saturating at the deuteron; Majorana's force increases the number of bonds to two, with each neutron able to bond to two protons and vice versa, naturally saturating at the α-particle. Heisenberg agreed with Majorana's results.

In a letter to Heisenberg of 14 July 1933, Pauli stressed the need to maintain the conservation laws of energy, momentum and angular momentum in nuclear processes (Pauli, 1985). So, for example, under no circumstances can a neutron with spin $\frac{1}{2}$ decay into a proton and an electron. As 'support for the validity' of the conservation law of angular momentum for nuclei, Pauli offered the stability of the hydrogen atom: Under no circumstances could the hydrogen atom be transformed into a neutron (with accompanying light emission). The only way that a neutron could decay into an electron and a proton, continued Pauli, is with an accompanying neutrino. Yet maintaining his stance against the positron, Pauli asserted that the proton could decay into a neutron and a positron with 'integral spin (Elsasser)' (see Section 4.4).

Yet 'if one believes in the hole theory the positron must possess all the characteristics of the electron (except for opposite charge). Consequently, the decays of neutron and proton are complicated processes such as neutron goes into proton plus other particles of an as yet unknown sort'. And Pauli concluded that if the '*fundamental ideas of the hole theory are correct, I would expect that the Austauschkräfte postulated by Majorana and yourself between neutron and proton do not exist*, or, if they do exist, they have a completely other source than the "*Austausch*"' (emphasis in original).

Heisenberg replied on 17 July 1933 that even if a neutrino were present in β-decay and all electrons did have spin $\frac{1}{2}$, 'then, too, would the *Austauschkräfte* enter. As for Elsasser, I believe him not at all'.

5.3 Discussion at Solvay in 1933

In the discussion session at Solvay in 1933 following Heisenberg's lecture, Pauli gave his opinion that the law of conservation of electric charge, which is always upheld, is no more fundamental than conservation of energy and momentum. But the situation is rather confusing with respect to the status of the latter laws because whereas on the one hand their validity is doubted, on the other hand they are used to calculate characteristics of the β-decay spectrum and nuclear stability considerations. Pauli went on to discuss his proposal of a neutrino ('so named by Fermi') whose proper mass may well be zero.

Francis Perrin noted that the form of the β-decay spectrum demands zero mass for the neutrino.

In contrast to conjectures on the nature of the neutrino, Bohr offered the view that since measurement of the energy of a β-particle is a 'well-defined process, the question of energy balance is imposed necessarily .... It is a question of taste what point of view one prefers [that is, nonconservation or conservation of energy] . . . No one knows what surprises again await us'.[56]

Dirac opposed the notion of Bose electrons, but was not totally against the possibility that in certain situations electrons could be nuclear constituents. This would mean that there are three elementary nuclear particles: 'This number can seem perhaps large, but from this point of view two is already a large number'. The number of nuclear electrons would always be very small so that the nucleus could be considered to be 'essentially constituted of protons and neutrons'. Dirac preferred Majorana's interaction to Heisenberg's because it contained more the flavor of a quantum mechanical process in which wave functions overlap rather than an actual exchange of something:

> The interactions by exchange represent a type of action so entirely fundamental that it is not possible to describe them in classical theory [we assume here that Dirac is rejecting any 'intuitive' depiction of Bose electrons 'migrating']. If protons and neutrons are both elementary particles, it is logical that their interaction should be of this fundamental type [describable with wave functions].

Other sorts of exchange forces were proposed, in addition to Heisenberg's and Majorana's.[57] But conceptually it was Heisenberg's visualizability of a nuclear exchange force with something actually exchanged that turned out to be fruitful. It led to Enrico Fermi's 1933 theory of β-decay, whose success, wrote Heisenberg in (1935), was 'proof of the existence of exchange forces' in nuclei.

5.4 Enrico Fermi's theory of β-decay

In (1934) Fermi solved the problem of the origin of the β-decay electrons with a theory of β-decay based on Pauli's neutrino, Heisenberg's isotopic spin formalism and second quantization.

Fermi wrote that in order to understand β-emission one should seek 'a theory of emission of light particles from a nucleus in analogy to the theory of the emission of a light quantum from a decaying atom according to the usual basis of radiation processes'. Fermi was able to reap the benefits of Heisenberg's 1932 *Anschaulichkeit* by viewing the exchange process differently from that shown in Figure 5.1(*g*). Rather, for β-decay, Fermi considered the exchange analogous 'to the emission of a light quantum' in a two-level atomic system.

Here, both neutron and proton were considered to be as elementary as the energy levels in an atom. So, according to Fermi, just as the light quantum emitted by an atom in an atomic transition was not present in the atom prior to the transition, neither was the electron present in the nucleus prior to β-decay.

Fermi's interaction Hamiltonian in occupation number space is

$$H_{\text{int}} = g \int (\psi_{\text{p}}^{+} \psi_{\text{n}})(\psi_{\text{e}^-}^{+} \psi_{\nu})\, \mathrm{d}^3 r, \tag{5.7}$$

where the integral is carried out over the coordinates of the neutron and Fermi estimated the coupling constant to be $g = 4 \times 10^{-50}$ erg cm^{-3}.

In his 1936 book *Anschauliche Quantentheorie*, Jordan wrote elatedly of Fermi's '*anschaulich* thinking of the analogy between γ-radiation and β-radiation'.

5.5 β-Decay and the nuclear force

On 18 January 1934 Heisenberg described to Pauli how in second-order perturbation theory Fermi's interaction in Eq. (5.7) produces an *Austauschwirkung* between neutron and proton by exchange of an electron and neutrino (Pauli, 1985). His result is

$$J(r) \sim \frac{g^2}{\hbar c}\frac{1}{r^5}, \tag{5.8}$$

which for $r \leq h/mc$ is too small to be the nuclear binding force. Thus, contrary to the hopes of Heisenberg and Fermi, a theory of β-decay could not cover the neutron–proton force as well.[58]

We note that Heisenberg used the term *Austausch* instead of *Platzwechsel*, thereby diluting somewhat the metaphor of forces that are transmitted by particle exchange. But Hideki Yukawa did not miss the intent of Heisenberg's original change of terminology. In November 1934 Yukawa considered the exchange force between the neutron and proton to be a 'migration force describable with a *Platzwechselintegral*' (Yukawa, 1935).

5.6 Hideki Yukawa's theory of the nuclear force: metaphor becomes physical reality

Yukawa set out to 'modify the theory of Heisenberg and Fermi' by changing the analogy of the field of 'light particles' transmitting the neutron–proton force, to the exchange of a single new particle between the neutron and the proton. He suggested that the transition of a 'neutron state to a proton state' need not always proceed through emission of a proton and electron,

> but the energy liberated by the transition is taken up sometimes by another heavy particle, which in turn will be transformed from proton state into neutron state . . . . Now such interaction between the elementary particles can be described by means of a field of force just as the interaction between the charged particle is described by the electromagnetic field.

Yukawa's Hamiltonian for the system of neutron, proton and the 'U-field', that is the interaction force between neutron and proton, is a straightforward generalization of the one in Heisenberg's 1932 isotopic spin paper where, as Yukawa wrote, 'we take for the "*Platzwechselintegral*"

$$J(r) = -g^2 \frac{e^{-\lambda r}}{r} \quad (5.9)$$

instead of the expression for $J(r)$ like the one in Eq. (5.1); in Eq. (5.9) g is the coupling constant for the interaction between neutron, proton and U-particle. At that time Yukawa ignored the interaction between neutrons and the Coulomb interaction between the protons. The quantity λ is directly proportional to the mass of the exchanged particle and inversely proportional to the range of the strong interaction force between neutron and proton.[59,60]

Is it not the case that Yukawa's explicit use of the German word *Platzwechsel* was to signal that he meant not a metaphorical exchange or *Austausch*, but a real 'migration'? Yukawa realized that the migrating particles must be Bose particles, which can be either positively or negatively charged, and have a mass about 200 times that of the electron. He offered the theory with frank reservations because 'such a quantum with large mass and positive or negative charge has never been found by experiment, [and therefore] the above theory seems to be on a wrong line'. He did, however, indicate that the new quanta 'may have some bearing on the shower produced by cosmic rays'.[61]

Despite proper *descriptions* of particles transmitting forces, nowhere in the scientific literature of the 1930s have I found *depictions* of this process.[62] The only depiction of β-decay that I found is in Figure 5.1(h), which is the sort of diagram that was the basis for Fermi's *anschaulich* β-decay theory.

In (1943) Wentzel proposed the 'didactic' device in Figure 5.1(i) for depicting Yukawa's treatment of β-decay as an 'indirect' process mediated by a virtual meson. The impact on Feynman of Wentzel's (1943) book remains to be ascertained owing to the complexity of the route that led Feynman to his diagrammatic methods.[63]

In his (1965) Nobel Prize address Feynman wrote that the absence of rigorous mathematical proofs for his diagrammatic rules caused the 'work [to be] criticized, I don't know whether favorably or unfavorably, and the "method" was called the "intuitive method"'.

Heisenberg's approval in (1950) of Feynman's diagrams as intuitive methods (*anschauliche Methoden*) (see the Epilogue) agreed with the new meaning of *Anschaulichkeit* in quantum electrodynamics. For example, the ordinary intuition (*Anschauung*) of the Coulomb force between two electrons is in Figure 5.1(e), which is abstracted from phenomena that we have actually witnessed in the world of sense perceptions. Such a drawing is *imposed* upon classical physics. The Feynman diagram in Figure 5.1(j) *emerges* from the mathematical formalism of quantum electrodynamics, and we would not have known how to draw it without this formalism. In Section 1.3 we discussed how the work of Bohr, Kramers and Slater initiated movement toward permitting visualizability to emerge from the mathematics of a physical theory. In this case, as it would be until 1932, visualizability did not have a visual component (see Section 1.9).

In contrast to Feynman's diagrams is the formalism of Schwinger and Tomonaga, which is the fruit of nonimaginal creative thought. Just as in many cases of productive thinking, problem situations attain a higher plateau of clarification once the proper diagram is drawn. Consequently it is not surprising that most physicists are more comfortable with Feynman's formalism.

Epilogue

Heisenberg's research style was to seek a new theory through correspondence-limit procedures to extend the concept of intuition into ever-smaller spatial domains. This approach had been incredibly successful in 1925 (invention of quantum mechanics), 1932 (invention of exchange forces in nuclear theory), and in 1934 it provided optimism (density-matrix formalism). Developments of Heisenberg's 1943 S-matrix formalism, particularly by Dyson (1949a,b), would lead to far-reaching results.

By 1943 Heisenberg judged the situation in physics as serious enough to warrant return to his strategy of 1925 in combination with the 'fundamental length' (Heisenberg 1943a). He proposed to replace the existing quantum electrodynamics with the S-matrix formalism based on only measurable quantities like cross-sections that could be calculated from the initial and final states of scattering processes. He hoped that the S-matrix could provide the means to penetrate interaction distances less than the fundamental length.

In Heisenberg's opinion the crux of basic problems in elementary particle physics was the Hamiltonians from classical physics which were essentially for point particles. We have seen this opinion surface in Heisenberg's letters of 12 March 1934 to Bohr (Section 4.7.3), and 5 February 1934 to Pauli, where Heisenberg proposed a new Hamiltonian that contains 'measurable quantities'. It is also part of the view of Heisenberg and Pauli which appears from time to time according to which correspondence-limit arguments between quantum electrodynamics, quantum mechanics and classical physics fail (for example, letters of Pauli to Peierls, 18 June 1929 in Section 3.2 and Heisenberg, 1930, analyzed in Section 4.2). Heisenberg expressed the hope that a future theory would contain the interaction Hamiltonian in a new way that would go over in a correspondence limit dependent on the fundamental length to the interaction Hamiltonian from classical physics (Heisenberg, 1943b, 1946; see, too, Rechenberg, 1989, and Cushing, 1990). Recalling problems of interpretation in 1927 for quantum mechanics, Heisenberg wrote (1943b) that the next step after formulating the new theory of elementary particles is to consider its 'intuitive contents [*anschaulichen Inhalt*]'.

It turned out that the technical apparatus required for setting up a method to cancel three of the four possible infinities in quantum electrodynamics existed already in the 1930s. These infinities are the electron's self-energy, vertex renormalization, wave renormalization, and vacuum polarization.

Removal of the divergent self-energy of the electron is accomplished with a hybrid version of the subtraction physics. The electron's *finite* measured mass m_{meas} is used in the H_{int} from the Heisenberg–Pauli quantum electrodynamics (Eq. (3.22)) which must be rewritten as

$$H_{\text{int}} = -\text{ie}\ \bar{\psi}\gamma_{\mu}\psi A^{\mu} - \delta mc^2\bar{\psi}\psi. \tag{E.1}$$

The quantity δm takes into account that the electron's measured mass

$$m_{\text{meas}} = m_{\text{bare}} + \delta m \tag{E.2}$$

is used in all calculations, where m_{bare} is the mass of the electron when it is not accompanied by its electromagnetic field and δm is the infinite contribution from the interaction of the electron with its self-field. The quantity $-\delta mc^2\bar{\psi}\psi$ serves to subtract the infinite contribution to the electron's mass so that quantum electrodynamics deals with a 'dressed' instead of a 'bare' electron. (Eqs. (4.61) and (E.2) are analogous.) But in order to set this program into motion the quantities in Eq. (E.1) have to be expressed in the interaction representation, which had already made its appearance in Dirac's many-time formalism (Dirac, 1932, and Dirac, Fock and Podolsky, 1932).

Two other divergent quantities cancel each other: vertex renormalization cancels wave function renormalization. Only charge renormalization remains, which is used to define the measured charge of the electron analogously to the manner suggested by Weisskopf (1936) (see Eq. (4.56) and note 40).

The concept of a fundamental length proved unnecessary in renormalized quantum electrodynamics. But it was important for Heisenberg's invention of the S-matrix formalism which reached its fullest potential upon jettisoning the concept. So Heisenberg was partly correct in his conjecture to Bohr that 'we are using an incorrect Hamiltonian' (12 March 1934 letter to Bohr in Section 4.7.3).

With the available mathematical methods and physical concepts what could Heisenberg have accomplished successfully by 1943? In order to maintain covariance at every stage of a calculation (see note 21) he had to combine proper development of the interaction representation (accomplished by Tomonaga in 1946) with the concept of renormalization. This concept was already present in subtraction physics, in Weisskopf's conjecture in [8] to replace subtraction with extraction in vacuum polarization and articulated by Kramers in [11] as the prescription to use right from the start measured masses in the Hamiltonian. (We recall that the desire to deal only with measurable quantities in quantum mechanics goes back to Kramers, 1924, see Section 1.4, and was expressed early on in quantum electrodynamics by Dirac, 1933; see Section 4.5.) In turn, all of this had to be assembled in a version of S-matrix formalism that contains H_{int} from Eq. (E.1) and uses a normal product ordering that

Heisenberg had been advocating since 1927 (see Section 2.3). These steps would have required Heisenberg's 'returning' to Heligoland to convince himself of the importance of using measurable quantities in Hamiltonians. In this scenario new data, such as the Lamb shift, play a secondary role, at best.

Question: Why did this scenario not come to pass? *Replies*: (1) Heisenberg's opinion that physical theories should not contain infinite quantities delayed any thoughts he had along the lines of a renormalized quantum electrodynamics. A 'return' to Heligoland would have diluted this opinion. (2) The war intervened, cutting off channels of scientific communication, which broke the continuity in Heisenberg's thoughts on quantum electrodynamics, thoughts that required Pauli as a sounding board just like in 1924–7 (see Miller, 1984, 1990, pp. 3–15). As Pauli wrote (in English) to Nicholas Kemmer on 24 November 1939 (Pauli, 1985):

> [O]ur famous colleague H[eisenberg] in Germany, with whom I had a regular exchange of letters about physics until to the End of August wrote a letter to Wentzel, that seemed to indicate that he doesn't venture more to write me, though he would like to do so very much himself . . ..

Issues other than quantum electrodynamics concerned physicists during 1939–45.

During 1946–8 the new quantum electrodynamics emerged in two apparently unrelated forms. On the one hand there was the formalism of Sin-itoro Tomonaga and Julian Schwinger with an aura of mathematical rigor. On the other hand there was Richard P. Feynman's 'intuitive' formalism utilizing diagrams based on mathematical rules seemingly lacking in rigor (see Section 5.6). In 1949 Dyson demonstrated that the key concept uniting these two approaches was Heisenberg's S-matrix.[64,65]

Dyson wrote (1949b): 'The Feynman method is essentially a set of rules for the calculation of the elements of the Heisenberg S-matrix corresponding to any physical process'. But just like in the 'subtraction physics', in Eq. (E.1) there is subtraction between two infinities. Thus, Dyson wrote that 'paradoxically opposed to the finiteness of the S-matrix is [that] the whole theory is built upon a Hamiltonian formalism with an interaction-function [Eq. (E.1)] which is infinite and therefore physically meaningless'.

This was not satisfactory, and Dyson tried to come to terms with the situation as a contrast between two observers. The tone of his methodology should ring familiar from the 1930s: There is the 'ideal' observer whose 'picture is of a collection of quantized fields with localizable interactions' who makes the sort of observations employed by Bohr and Rosenfeld: This observer uses apparatus with no atomic structure and capable of an accuracy limited by the uncertainty relations. Then there is the 'real' observer whose 'picture is of a collection of observable quantities (in the terminology of Heisenberg) [with] apparatus that consists of atoms and elementary particles [and who] makes spectroscopic observations and performs' scattering experiments. The ideal observer uses the Hamiltonian formalism and the idealized measurements of Bohr and Rosenfeld. The real observer uses the S-matrix and cannot measure the strength of a single field. So far so good. The 'paradox is the fact that it is necessary [for the real observer] to start from the infinite expressions in order to deduce the finite ones'. In the future theory the situation would be just the opposite: 'finite quantities will become primary and the infinite quantities secondary'.

Contrary to the programmatic intent of Heisenberg and Pauli, who sought to cure the ills of quantum electrodynamics through new formulations with experimental data playing a secondary role, Dyson proposed the opposite route: 'the future theory will be built, first of all upon the results of future experiments'. In further opposition to the views of Heisenberg, Dyson concluded that there was no longer 'a compelling necessity for a future theory to abandon some essential features of the present electrodynamics. The present electrodynamics is certainly incomplete, but is no longer certainly incorrect'.

The new results, in fact, suited Heisenberg's taste in theoretical physics. He wrote (1950):

> The researches of Schwinger and Dyson on quantum electrodynamics have taught us how we deal with nonlinear reciprocal effect terms, and how the correspondence limit-*anschaulich* representation can be effected after carrying out certain renormalizations. With it has been also established the connection to the S-matrix formalism that had been developed independently of quantum electrodynamics.[66]

Once again for Heisenberg theory decided what is intuitive: Feynman diagrams were the new *Anschaulichkeit*. For although Feynman diagrams are drawn with the usual distinction between figure and ground, without the mathematics of quantum electrodynamics we would not have known how to draw these diagrams at all. As Heisenberg recalled of his own research: 'The picture changes over and over again, it's so nice to see how such pictures change' (*AHQP*: 11 February 1963).

Notes

1 The correspondence principle states that in the region of high quantum numbers, where an atom's spectral lines are so closely spaced as to be almost a blur, quantum mechanics and classical electromagnetic theory asymptotically approach each other. Bohr's procedure goes as follows (Bohr, 1918): For transitions between orbits whose principal quantum number $n \gg 1$ the quantum frequency of the emitted radiation ν_{q} is nearly equal to the classical frequency ν_{c} of the electron's revolution in either of these orbits, that is

$$\nu_{\mathrm{q}} = \nu_{\mathrm{c}} = \sum_k \tau_k \nu_k, \tag{A}$$

where the classical frequency is the sum of higher harmonics and the τ_k are integers.

For the purpose of studying the response of atoms to radiation, classical electrodynamics represents the atom as comprising of harmonically bound electrons. The atom's dipole moment is

$$P(t) = \sum_{\tau_i} C_{\tau 1 \ldots \tau s} \exp\{2\pi \mathrm{i}(w_1 + \cdots + w_s)\}, \tag{B}$$

where $w_i = \nu_i t$ are angle variables and the coefficients of the Fourier expansion $C_{\tau 1 \ldots \tau s}$ are functions of action variables J_i. But the spectrum of emitted radiation from an atom predicted by classical theory (as equally spaced spectral lines) differs completely from the one measured and the one predicted by Bohr's atomic theory.

According to classical physics, the rate at which an atom emits radiation is deduced from Eq. (B) as

$$\frac{\mathrm{d}E_{\tau 1 \ldots \tau s}}{\mathrm{d}t} = \frac{(2\pi\nu_{\mathrm{c}})^4}{3c^3}|C_{\tau 1 \ldots \tau s}|^2. \tag{C}$$

Bohr drew upon the correspondence principle (Eq. (A)) to rewrite Eq. (B) as

$$P(t) = \sum_q U_q \exp\{2\pi \mathrm{i} \nu_q t\}, \tag{D}$$

where the summation is over all quantum jumps q, and then to express the rate at which radiation is emitted for a spontaneous transition between stationary states i and k as

$$\frac{\mathrm{d}E_q}{\mathrm{d}t} = \frac{(2\pi\nu_q)^4}{3c^3} e^2 |C_q|^2. \tag{E}$$

For further discussion of Bohr's correspondence principle, see Jammer (1966) and Miller (1990, pp. 3–15).

2 Dirac's derivation (1926a) goes thus: Since all electrons are identical particles, then the state (m, n) is the same as the state (n, m). Consequently the transitions $(m, n) \to (m', n')$ cannot be distinguished from $(m, n) \to (n', m')$. In the absence of interactions the eigenfunction for a two-electron system is

$$\psi_{nm} = a\psi_n(1)\psi_m(2) + b\psi_n(2)\psi_m(1).$$

If Q is an operator symmetric in the coordinates of the two electrons, then Q can be represented in matrix elements in the space spanned by ψ_{nm} as

$$Q\psi_{nm} = \sum_{n'm'} \psi_{n'm'} Q_{n'm',nm}$$

only if $a = \pm b$.

3 On 22 November 1926 Heisenberg reported to Pauli (1979) that Dirac 'has managed an extremely broad generalization of my fluctuation paper.' For a discussion of this paper see Miller (1990, pp. 3–15).

4 Hartree (1928) used the correspondence principle to argue that for a many electron system one could impose the condition $\int \psi_k^* \psi_k \, d^3r = n_k$ for each mode of vibration k of the system.

5 In (1948) H. B. G. Casimir found that this was not true in general, but only for the vacuum.

6 The canonical transformation is:

$$P = \frac{1}{2(\pi v m)^{1/2}} p - i(\pi v m q)^{1/2}$$

$$Q = \frac{1}{2(\pi v m)^{1/2}} p + i(\pi v m q)^{1/2},$$

where

$$[P, Q] = i[p, q] = \hbar,$$

then

$$H(P, Q) = 2\pi v Q P = E,$$

and

$$E_n = nhv,$$

where in this case the eigenfunction $\psi_E(Q)$ is a complex quantity.

7 This equation was proposed also by Schrödinger in (1926b). For further discussion see Kragh (1981).

8 See, too, Waller (1930a), where this point was rediscovered. In a footnote Waller wrote that Pauli had informed him of Heisenberg's unpublished calculation in Heisenberg's letter to Pauli of 31 July 1928.

9 Other 'scandals' were also occurring. In (1928) G. Breit found that for a free electron the eigenvalue for α_k is $\alpha_k = \pm 1$, which means that the eigenvalue of the velocity operator, which is $dx_k/dt = c\alpha_k = \pm c$. But there is no problem here, reasoned Breit, because unless the velocity of the electron is zero the plane-wave solutions of the Dirac equation are not simultaneously eigenvalues of momentum and α_k. In addition, since α_k does not commute with the Hamiltonian operator

$$H = -i\hbar c \boldsymbol{\alpha} \cdot \boldsymbol{\nabla} + mc^2 \beta$$

(unless $v = 0$) then neither are energy eigenfunctions simultaneous eigenfunctions of α_k.

10 Heisenberg and Pauli proved the relativistic invariance of the commutators and anticommutators in their field quantization procedure.

11 Working to achieve relativistic invariance, in Part I, Heisenberg and Pauli (1929) used the Lorentz gauge instead of the Coulomb gauge used by Dirac in (1927a).

12 Heisenberg had realized this point a year earlier (letter of Pauli to Dirac 17 February 1928 in Pauli, 1985).

13 Heisenberg and Pauli cite Klein's paradoxical result to be 'especially striking' because it emphasized the negative energy difficulty and 'seems to thwart detailed treatment of nuclear structure'. The reason is that the sort of potential well simulating the nucleus would permit electrons to escape from the nucleus with negative kinetic energies.

14 Moreover, they continued, quantization of the gravitational field 'should be feasible without any new difficulties by means of a formalism completely analogous to the one used here'.

15 Oppenheimer (1930b) also showed that, in the limit $v/c \ll 1$, Heisenberg and Pauli's iteration method in Part II becomes the one for a Schrödinger equation in configuration space for the many-body problem.

16 The argument leading to Eq. (3.33) follows Heitler (1936). Concerning Eq. (3.34), Heitler noted that quantum electrodynamics produces satisfactory results for the energy loss of cosmic rays in matter up to 150 mc^2 and then breaks down. See Section 4.14 and note 41 for further discussion of cosmic ray physics in the 1930s.

17 In a letter to Bohr (26 November 1929) Dirac speculated that in the absence of electromagnetic interactions the masses of the electron and the proton are equal. Quoted in Pais (1986).

18 Most likely they had in mind the following two points:

(1) A localized free particle wave packet contains negative energy states which become more predominant with increased localization. For a packet localized to within the electron's Compton wavelength, negative energy states become the predominant component.

(2) The '*Zitterbewegung*' phenomenon discovered by Schrödinger in (1930). Using the representation where operators are time dependent and wave functions are time independent (which became known as the Heisenberg picture), Schrödinger calculated the velocity of the electron to be

$$\frac{\mathrm{d}x_k}{\mathrm{d}t} = \frac{\mathrm{i}}{\hbar}[H, x_k] = c\alpha_k. \tag{A}$$

From Eq. (3.9) the operator Hamiltonian for a free electron is

$$H = -\mathrm{i}\hbar c\boldsymbol{\alpha}\cdot\boldsymbol{\nabla} + mc^2\beta, \tag{B}$$

and so

$$\frac{\mathrm{d}x_k}{\mathrm{d}t} = \pm c \quad \left(\text{expectation value of } \frac{\mathrm{d}x_k}{\mathrm{d}t}\right). \tag{C}$$

Schrödinger continued, 'one wonders how the center of gravity of the charge cloud [that is, the electron's wave packet] can manage to move so fast . . .. Evidently, now, by not moving in a straight line'.

In order to investigate this non-straight-line motion Schrödinger calculated

$$\frac{\mathrm{d}\alpha_k}{\mathrm{d}t} = \frac{\mathrm{i}}{\hbar}[H, \alpha_k] = \frac{1}{\hbar}(-2\alpha_k H + 2cp_k) \tag{D}$$

then

$$\frac{\mathrm{d}x_k}{\mathrm{d}t} = \frac{1}{\hbar}(-2x_k H + 2c^2 p_k), \tag{E}$$

and

$$x_k = x_k(0) + c^2 p_k H^{-1} + \frac{\mathrm{i}c\hbar}{2}(\alpha_k(0) - cp_k H^{-1})H^{-1}\exp(-2\mathrm{i}Ht/\hbar). \tag{F}$$

The first two terms in Eq. (F) are expected from classical physics. The last term is new. It gives the '*Zitterbewegung*' or trembling motion of the wave packet which is superposed on the straight line motion. *Zitterbewegung* arises from the interference terms between negative and positive energy components of the wave packet. A simple calculation shows the fluctuation to be of the order of the Compton wavelength of the electron

(3×10^{-13} m). Schrödinger interpreted $c\alpha_k$ to be the 'instantaneous velocity of the center of gravity of the charge cloud' or the 'electron's microscopic velocity'. The electron's measured velocity is $cH^{-1}p_k$, which is always less than c. The root of these apparent incompatibilities is that the α_k do not commute with the Hamiltonian – see, for example, Eq. (B), which is for a free electron.

19 Heisenberg was the first to explore limits to electromagnetic field measurements (Heisenberg, 1930b). Bohr and Rosenfeld (1933) showed that Heisenberg's results hold only for measurements in uninteresting space-time domains. For historical analysis of the Bohr–Rosenfeld paper see Miller (1990, pp. 139–52), and for further developments of the Bohr–Rosenfeld results see Corinaldisi (1953).

20 Pauli's own *raison d'être* for the $\Delta^{1/2}$ was that new devices were needed for systems with an infinite number of degrees of freedom, such as the radiation field. The reason is that even in the limit of large quantum numbers the radiation field has an infinite number of degrees of freedom: 'Use of the wave mechanical formalism appears, therefore, to be justified on grounds of the correspondence principle if we confine ourselves to a finite number of degrees of freedom' (Pauli, 1933). That is, in k-space, high k-values are disregarded, and in configuration space only the averages of field strengths are considered. For further discussion see Landau and Peierls (1930) and Pauli's *Handbuch* (1933). The problem of the electromagnetic field's zero-point energy was always on Pauli's mind (see C. P. Enz's introduction to the translation of Pauli's *Handbuch* article, 1933).

21 Although the Heisenberg–Pauli quantum electrodynamics and the Heisenberg density-matrix formalism (to be discussed in Section 4.7.4) are initially relativistic and gauge invariant, relativistic invariance is not maintained throughout calculations (for example, initial and final times are set equal) and the calculations are done in a particular gauge, the Coulomb gauge. As Schwinger (1958) wrote, 'in virtue of the divergences inherent in the theory, the use of a particular coordinate system or gauge in the course of computation could result in a loss of covariance. A version of the theory [quantum electrodynamics] was needed that manifested covariance at every stage of the calculation'.

In (1946) Sin-itoro Tomonaga realized that the many-time formalism was precisely the one needed to ensure relativistic covariance at every stage of a calculation. Tomonaga's generalization of the many-time formalism, along with Schwinger's work done simultaneously but independently, is one of the paths to modern renormalization theory. The other one is the less formal, and so more intuitive, path taken by Feynman. For discussion on the background of Tomonaga's work, which he began in 1944, see Brown *et al.* (1988) and Darrigol (1988b).

22 Heisenberg and Pauli (Heisenberg, 1930a) had broached the subject of a coordinate space representation, which was developed further by Oppenheimer (1930b).

23 See Heitler (1936, p. 196) and discussion session at Solvay of Dirac's 1933 paper reprinted here with discussion session as [3].

24 In modern notation Heisenberg calculated the vacuum expectation value of the charge squared, that is the charge fluctuations

$$\langle e^2 \rangle = \left\langle \int\int \rho(x')\rho(x'')\,\mathrm{d}^4x'\,\mathrm{d}^4x'' \right\rangle .$$

For further discussion of Heisenberg's calculation see Oppenheimer (1935) and Corinaldisi (1953) and for the relation of Heisenberg's results to the field fluctuations in Bohr and Rosenfeld (1933).

25 This replaces the too-complicated approach of a single wave function that is dependent on the variables of every electron present. In the Hartree method each electron has its own wave function, and the assumption is that each electron moves in an effective field

that is the same for all electrons. In fact, it was in order to remove the arbitrariness in the way that a totally antisymmetric wave function for n electrons is constructed that Dirac invented the density matrix (Dirac, 1930c).

26 I am grateful to Prof. Weisskopf for showing me this hitherto unpublished letter. He has described this episode in his career in Weisskopf (1989): 'Furry was a gentleman. [He insisted that] I should publish a correction in my name only and mention him as the person who drew my attention to the error. Since then, the logarithmic divergence of the self-energy of the electron goes with my name and not with Furry's'.

27 In (1939) Weisskopf showed that to order α^n the electron's self-energy diverges like $(\log(P/mc))^n$, that is the 'divergence is logarithmic in every approximation'.

28 I am grateful to Prof. Weisskopf for permitting me to cite from this unpublished letter.

29 Oppenheimer and Furry (1934) used a method to calculate vacuum polarization that, for static fields, is equivalent to Heisenberg's.

30 Serber (1936) extended the vacuum polarization result to a time varying potential and obtained

$$\delta\rho(r, t) = \frac{1}{15\pi}\frac{e^2}{\hbar c}\left(\frac{\hbar}{mc}\right)^2 \Box\rho_0.$$

31 It turned out that Heisenberg's subtraction physics is required to calculate the small probability for light–light scattering at low frequencies (Euler, 1936).

32 In (1936) Serber discovered an error in Heisenberg's calculation. In the canonical transformation for second-order perturbation theory, Heisenberg had taken improper limits for quantities singular on the light cone. According to Serber's calculations, in the density-matrix formalism the photon should have a finite self-energy.

From time to time historians make special mention of Serber's use of the term 'renormalize'. He writes that the canonical transformation matrix in Heisenberg's application to second-order perturbation theory in Eq. (63)–[6] 'is chosen to renormalize the polarization of the vacuum for slowly varying and weak fields to zero'. Clearly Serber's term 'renormalize' is a substitute for the term subtraction. Weisskopf's statement about vacuum polarization as extraction of an infinite constant instead of subtraction (see Section 4.13) is much more far-reaching than Serber's 'renormalize' understood in the *context* of physics in 1936.

33 Weisskopf (1989) recalled that the Pauli–Weisskopf theory is a direct result of his own work on the properties of $|\psi|^2$ for the Klein–Gordon equation:

> I was struck by the fact that the wave intensity $|\psi|^2$ is not conserved in the presence of electromagnetic fields, whereas the expression for the charge density is different from $|\psi|^2$ and fulfills the charge conservation laws. I felt there might be something like a lack of conservation of particle number in that equation and that this might lead to pair creation or annihilation.

This result piqued Pauli's interest. See, too, Weisskopf's interviews in *AHQP* (10 July 1963).

34 In this letter Pauli described an attempt to force spinors into the energy-density term from the Klein–Gordon equation for a particle in a field with potential φ

$$\left(\frac{\hbar}{\mathrm{i}}\frac{\partial\psi^+}{\partial t} + e\varphi\psi^+\right)\beta\left(\frac{\hbar}{\mathrm{i}}\frac{\partial\psi}{\partial t} - e\varphi\psi\right) + \cdots,$$

which turned out to be as 'often positive as negative'.

35 Weisskopf (1989) recalled that Pauli referred to the Pauli–Weisskopf paper as the 'anti-Dirac paper'. Since Pauli and Weisskopf considered their theory to be truly a

'curiosity', they chose to publish it in the venerable but not first-line journal *Helvetica Physica Acta*. (Private communication from Prof. Weisskopf.)

36 The Lorentz invariant D-function is

$$D(x, t) = \frac{1}{(2\pi)^3} \int \mathrm{d}^3 k e^{i\mathbf{k}\cdot\mathbf{x}} \frac{\sin k_0 t}{k_0},$$

where $(\mathbf{k}, k_0)$ are components of a four-vector k_μ. The D-function satisfies the Klein–Gordon equation

$$(\Box - \kappa^2)D = 0$$

and vanishes outside of the light cone. For $\kappa = 0$ it becomes the Jordan–Pauli D-function in Eq. (2.61). The other possible Lorentz invariant quantization function that satisfies the Klein–Gordon equation is the D_1-function, which does not vanish outside of the light cone. The Lorentz invariant D_1-function is

$$D_1(x, t) = \frac{1}{(2\pi)^3} \int \mathrm{d}^3 k e^{i\mathbf{k}\cdot\mathbf{x}} \frac{\cos k_0 t}{k_0}.$$

The analytic properties in coordinate space for the D- and D_1-functions are given explicitly in Eqs. (19) and (22), respectively, in [4].

37 Kemmer and Weisskopf (1936) calculated light–light scattering with an external electrostatic field (Delbrück scattering) without using subtraction physics which they shunned 'because of arbitrariness'.

38 Heisenberg and Euler (1936) briefly discussed the similarity of their phenomenological Lagrangian for light–light scattering with the one from the Born–Infeld theory of the electron. Born and Leopold Infeld made the 'modified Maxwell equations the starting point for their theory', whereas in quantum electrodynamics modification of electromagnetic theory is an 'indirect consequence of the virtual possibility of pair production'. Owing to the then problems of quantum electrodynamics Heisenberg and Euler did not discount the Born–Infeld theory.

Owing to difficulties in quantizing it, interest faded in the Born–Infeld theory. As Pryce (1937) wrote, the Born–Infeld theory has 'no mechanism for the production and annihilation of pairs'. For a survey of the Born–Infeld theory see Heitler (1936).

39 Weisskopf reached this conclusion by analogy with a charged body placed into a polarizable medium. The measured charge is the original charge reduced by the medium's polarizability. (Private communication from Prof. Weisskopf.)

40 For completeness, Eq. (4.46) should be written

$$e_1 = Z_1^{-1} Z_2 Z_3^{1/2} e,$$

where Z_1 (Z_2) is the infinite constant from wave function (vertex) renormalization. It turns out that to all orders in α, $Z_1 = Z_2$, and so Z_3 is interpreted as a charge renormalization. That vacuum polarization is the only correction to the electron's charge is important because Z_3 is independent of the type of Dirac particle. Eq. (4.56) holds, for example, for muons too. For further discussion see Dyson (1949a,b, 1951a,b), Schweber (1961) and Sakurai (1967).

41 For cosmic ray showers Heisenberg used the terms 'explosions' and 'cascades'. An explosion is when a high energy cosmic ray proton collides with a nucleus and a many-body final state emerges all at once. In contrast, a cascade is a gradual build up from particles produced by *Bremsstrahlung* and pair creation.

Until 1936 data did not agree with quantum electrodynamics at particle energies in excess of $150mc^2$ (see note 16, and recall that beyond $137mc^2$ quantum electrodynamics was expected to break down). In 1936 improved cosmic ray data separated into soft and hard components revealed that every shower is a cascade. The soft component could be dealt with by quantum electrodynamics to energies well beyond $137mc^2$. Analysis of the hard component revealed that either quantum electrodynamics fails or there are particles with mass intermediate between those of the electron and the proton. This possibly new cosmic ray particle was thought to be Yukawa's meson. In 1947 empirical data showed that the 1937 meson was not Yukawa's but the weakly interacting μ-meson. During the 1930s full blame for problems in interpreting cosmic ray data was placed on theory, physicists being reluctant to postulate new particles. For discussions of cosmic ray physics in the 1930s see Cassidy (1981, 1992), Pais (1986) and Galison (1987).

42 The coupling constant that Heisenberg used in the perturbation calculation is

$$f = \frac{g}{\hbar c} \sim 10^{-31}\ \text{cm}^2$$

in analogy to the expansion parameter in quantum electrodynamics, which is the dimensionless fine structure constant $\alpha = e^2/\hbar c$. From dimensional analysis Heisenberg found that the cross-sections depended on the quantity $f^{1/2}/\lambda$ where λ is the wavelength corresponding to the interaction energy between the incident proton and the nucleus. Consequently, for high interaction energies $\lambda < f^{1/2}$, the cross-sections become increasingly divergent and the number of particles produced increases.

43 Reference here to Heisenberg's invention of quantum mechanics during a May 1925 sojourn on the island of Heligoland where he went to recover from a severe attack of hay fever.

44 Heisenberg digressed to discuss the 'problem of gravitational forces' and noted here too that there is a fundamental length

$$L = (\hbar\gamma/c^3)^{1/2} = 4 \times 10^{-35}\ \text{m},$$

where γ is the gravitational constant. Since $L \ll r_0$, then the 'nonintuitive features of gravitation' can be neglected for studies in atomic physics.

45 J. J. Sakurai (1967) put it well when he wrote that 'if we ask the "right" question, the so-called infrared catastrophe is *not* a catastrophe at all' (emphasis in original). The 'right' question is: What exactly is measured in the scattering of an electron by an external potential? Reply: The cross-section for elastic scattering and the cross-section for emission of soft photons that are perhaps undetectable, depending on the detector's resolution. If the detector's resolution is ΔE, then photons with energies less than ΔE are not counted. In any case the measured finite scattering differential cross-section becomes independent of the cutoff frequency when it is written as the sum of an elastic part and a *Bremsstrahlung* part.

Another incredible cancellation of a covariant S-matrix expansion for the scattering of an electron by a quantized radiation field is that to any order in α the infrared divergence vanishes when added to the cross-section for radiative corrections with no emission of a real photon. For further discussion see Schweber (1961) and Sakurai (1967).

46 The β-ray spectrum's continuous nature (in velocity) had been known since 1902 when it was considered to be merely an inconvenience for measuring the charge-to-mass ratio of high-velocity electrons (see Miller, 1981). By 1927 the spectrum's continuity had been established as a property of the disintegrating nucleus (see Pais, 1986).

47 Before the discovery of the neutron in 1932, the nucleus was assumed to be composed of electrons and protons. If the N^{14} nucleus were composed of protons and electrons, then it would have fourteen protons and seven electrons and so would obey Fermi–Dirac statistics. But band spectra indicated Bose statistics. The most straightforward way out of this impasse was the proposal that nuclear electrons do not affect the statistics of the nucleus. Therefore, nuclear electrons do not obey quantum theory, Dirac's equation included because their spins are suppressed, among other reasons. Interpretational

problems of the negative energy states instilled doubt that Dirac's theory could be applied to particles such as nuclear electrons. Moreover, according to the uncertainty principle, nuclear electrons should have energies in excess of $137mc^2$, which is the region where quantum electrodynamics was expected to break down. Yet another problem with an electron–proton model of the nucleus was Klein's paradox, according to which a nuclear electron could escape by changing the sign of its mass. (For further discussion see Gamow, 1931.)

48 The conception of the neutron as a bound state, or as a collapsed hydrogen atom, had been proposed by Rutherford in (1920) and advocated by the discoverer of the neutron James Chadwick.

49 Another choice for heavier nuclei was to assume that they were composed of fundamental neutrons, protons and electrons, where quantum mechanics applied to the fundamental protons and neutrons but not to the electrons. But nuclear electrons would have to be bound to neutrons, protons and α-particles, which gave rise to yet another problem: Nuclear electrons would transfer some of their energy to α-particles because of the strong binding between them. But the emitted α-particles have a definite energy and yet the nuclear electrons to which α-particles are tightly bound do not have a definite energy. For a survey of the experimental data on nuclear electrons during 1920–34 see Stuewer (1983).

50 In (1932) Chadwick had dismissed this problem with the comment that the only advantage 'to suppose that the neutron may be an elementary particle [is] the possibility of explaining the statistics of such nuclei as N^{14}'. In his Nobel Prize Address, Chadwick, in turn, dismissed with short shrift the possibility of a composite neutron because it 'assigns the wrong spin to the electron' (Chadwick, 1935).

51 Pauli openly discussed the neutrino at a meeting in 1931 at Pasadena, California (see [3]). In 1930 Pauli had proposed the existence of the neutrino in a presentation at a radioactivity conference in Tübingen of 4 December 1930 that began with 'Dear radioactive ladies and gentleman' (Pauli, 1985). He originally considered the neutrino to have a very small mass. (For further discussion see Brown, 1978, and Pais, 1986.)

52 In fact, the difficulty of understanding exactly what Heisenberg was attempting to do in 1932 is attested to by the inaccurate ways in which his work was described. For example, Wigner (1933) wrote that Heisenberg discussed the possibility that the 'only elementary particles are the proton and electron'. Others at this time also speculated that the nucleus was composed of neutrons and protons. See, for example, the short note of Iwanenko (1932) with no details.

53 It is noteworthy that of all Heisenberg's reminiscences on the history of physics in the 1920s and 1930s that I have read, those pertaining to his 1932 nuclear physics paper are the least trustworthy. For example, he recalled:

> . . . to keep an electron in the nucleus would be a dreadful affair from the ordinary point of view of quantum theory. Therefore, if actually there were electrons in the nucleus, then probably you would have to change everything again. So I was extremely happy to see that this was not necessary (*AHQP*: 12 July 1963).

In his nuclear theory papers of 1932–3 Heisenberg tried in every way to include electrons (Bose electrons) in the nucleus.

54 Charge independence, that is equality of the proton–proton, neutron–neutron and proton–neutron forces, would not be empirically found until 1936.

55 In fact, not until after charge independence of nuclear forces was discovered in 1936 was Heisenberg's 'ρ-spin coordinate' resurrected by Wigner (1937).

56 As Pais (1986) observed, 'energy nonconservation would no doubt have died a quiet unheralded death had it not been for a last flare-up from an unanticipated direction'. In a 1936 measurement of the Compton effect, R. S. Shankland claimed to have found violation of the conservation laws of energy and momentum. Probably in despair at the apparent collapse of quantum electrodynamics, Dirac welcomed Shankland's results, which he conjectured might require something along the lines of a Bohr–Kramers–Slater theory. This statement caused Bohr to publically defend the conservation laws. For further discussion of these interesting about-faces see Pais (1986).

57 In (1933) Wigner introduced a third sort of nuclear force. It was a purely short range central force of a form meant to explore the enormous difference between the binding energies of the deuteron and the α-particle (in 1932 measured to be seventeen times larger). Needless to say, it could not lead to saturation. Nevertheless, using ordinary quantum mechanics Wigner initiated what became the enormous enterprise of fitting data such as binding energies with potentials of various shapes (see, for example, Evans, 1955).

58 These results were subsequently published by Iwanenko (1934) and Tamm (1934), who were unaware of Heisenberg's unpublished speculations.

59 Wick (1938) showed that the relation between the mass of the carrier of the nuclear force and the range of the force in Yukawa's theory follows from Heisenberg's uncertainty principle and so is independent of perturbation theory considerations.

60 For further discussion of the history of meson theory see Brown (1981), Pais (1986) and Darrigol (1988a, b).

61 The scientific literature supports Wentzel's recollection that Yukawa's 'ingenious idea . . . was not received, wherever it became known, with immediate consent or sympathy' (Wentzel, 1973). This situation changed in 1937 when Yukawa's meson was declared to have been found in cosmic rays. However, this turned out not to be the nuclear field or strong interaction meson, that is the subsequently found π-meson, but the weakly interacting μ-meson.

62 What follows is an outline of analysis in Miller (1984, 1985, 1989) of the development of diagrammatic methods from Heisenberg through Yukawa and Feynman. See Miller (1991) for an analysis of the importance of metaphorical notions during the genesis of quantum field theory in the 1930s and 1940s. For other relevant discussions of Feynman diagrams see Harré (1986) and Schweber (1986b).

63 In (1949) Feynman cites Wentzel's (1943) book.

64 Dyson's proof became more transparent with the help of Wick's (1950) version of the normal product expansion, which is the full formulation of what Heisenberg referred to as the 'Klein–Jordan Trick'. See Section 2.3 and Dyson (1951a, b).

65 Certain of Feynman's results were anticipated by E. C. G. Stückelberg in the 1940s. Since Stückelberg's papers on quantum electrodynamics were not in the mainstream of developments in the 1930s (which is the focus of this book), and, as far as I know, had little or no effect on Feynman, I have chosen not to discuss them. For commentary, see Pais (1986) and, especially, Rechenberg (1989).

66 Heisenberg, of course, cites Dyson (1949b).

References to the Frame-setting essay

The papers listed below were of use in writing the Frame-setting essay. For a more complete listing that is periodically updated see the *Isis Critical Bibliography*.

Abraham, M. 1902. Prinzipien der Dynamik des Elektrons. *Physikalische Zeitschrift* 4: 57–603

Abraham, M. 1903. Prinzipien der Dynamik des Elektrons. *Annalen der Physik* 10: 105–79

Bethe, H. and Salpeter, E. 1957. *Quantum Mechanics of One- and Two-Electron Systems*. Berlin: Springer

Bloch, F. 1934. Die physikalische Bedeutung mehrerer Zeiten in der Quantenelektrodynamik. *Physikalische Zeitschrift der Sowjetunion* 5: 301–15

Bloch, F. and Nordsieck, A. 1937. Note on the continuous radiation field of the electron. *Physical Review* 52: 54–9. Reprinted in Schwinger (1958)

Bohr, N. 1913. On the constitution of atoms and molecules. *Philosophical Magazine* 26: 1–25, 476–502, 857–75

Bohr, N. 1918. On the quantum theory of line-spectra. *Det Kongelige Danske Videnskabernes Selskab. Mathematisk-fysiske Meddeleser* IV: 1–118. Reprinted in van der Waerden (1968)

Bohr, N. 1923. Über die Anwendung der Quantentheorie auf den Atombau. I. Die Grundpostulate der Quantentheorie. *Zeitschrift für Physik* 13: 117–65. Version published as, On the application of the quantum theory to atomic structure: Part I. Postulates of the theory. *Proceedings of the Cambridge Philosophical Society (Supplement)* (1924)

Bohr, N. 1925. Atomic theory and mechanics. *Nature (Supplement)*, pp. 845–62

Bohr, N. 1928. The quantum postulate and the recent development of atomic theory. Versions are in: *Atti del Congresso Internazionale dei Fisici*. Bologna: Zanichelli, pp. 565–98; *Nature (Supplement)*, pp. 580–90; *Die Naturwissenschaften* 15: 245–57

Bohr, N. 1932. Chemistry and the quantum theory of atomic constitution. *Journal of the Chemical Society* 135: 349–84. Text of the Faraday lecture, delivered 8 May 1930

Bohr, N., Kramers, H. and Slater, J. C. 1924. The quantum theory of radiation. *Philosophical Magazine* 47: 785–802. An almost identical version is Über der Quantentheorie der Strahlung. *Zeitschrift für Physik* 24: 69–87

Bohr, N., and Rosenfeld, L. 1933. Zur Frage der Messbarkeit der elektromagnetischen Feldgrössen. *Det Kongelige Danske Videnskabernes Selskab. Mathematisk-fysiske Meddeleser* 12: 1–65. Translated by A. Peterson in *Selected Papers of Léon Rosenfeld* (eds. R. S. Cohen and J. Stachel). Reidel: Dordrecht

Born, M. 1926. Zur Quantenmechanik der Stossvorgänge. *Zeitschrift für Physik* 37: 863–7

Born, M., Heisenberg, W. and Jordan, P. 1926. Zur Quantenmechanik. *Zeitschrift für Physik* 35: 557–615. Translated in van der Waerden (1968)

Breit, G. 1928. An interpretation of Dirac's theory of the electron. *Proceedings of the National Academy of Science* 14: 553–9

Brink, D. M. 1965. *Nuclear Forces*. Oxford: Pergamon

Brown, H. R. and Harré, R., eds. 1988. *Foundations of Quantum Field Theory*. Oxford University Press

Brown, L. 1978. The idea of the neutrino. *Physics Today* 32: 23–8

Brown, L. 1981. Yukawa's prediction of the meson. *Centaurus* 25: 71–132

Brown, L. 1991. The development of vector meson theory in Britain and Japan (1937–38). *British Journal for the History of Science* 24: 405–33

Brown, L., Kawabe, R., Konuma, M. and Maki, Z., eds. 1988. *Elementary Particle Theory in Japan, 1935–1960*. Kyoto: Yukawa Hall Archival Library

Brown, L. and Moyer, D. 1984. Lady or Tiger? – The Meitner–Hupfeld effect and Heisenberg's neutron theory. *American Journal of Physics* 52: 130–6

Casimir, H. B. G. 1948. On the attraction between two perfectly conducting plates. *Koninklijke Nederlandsce Akademie van Wetenschappen: Proceedings of the Section of Sciences* 51: 793–5

Cassidy, D. 1981. Cosmic ray showers, high energy physics, and quantum field theories: Programmatic interactions in the 1930's. *Historical Studies in the Physical Sciences* 12: 1–39

Cassidy, D. 1992. *Uncertainty: The Life and Science of Werner Heisenberg*. Freeman: New York

Chadwick, J. 1932. The existence of a neutron. *Proceedings of the Royal Society of London (A)* 136: 692–708

Chadwick, J. 1935. The neutron and its properties. In *Nobel Lectures in Physics: 1922 to 1941*. New York: North-Holland

Corinaldisi, E. 1953. Some aspects of the problem of measurability in quantum electrodynamics. *Il Nuovo Cimento (Supplemento)* 10: 83–100

Cushing, J. 1990. *Theory Construction and Selection in Modern Physics: S-matrix*. Cambridge University Press

Darrigol, O. 1986. The origin of quantized matter waves. *Historical Studies in the Physical Sciences* 16: 197–253

Darrigol, O. 1988a. The quantum electrodynamical analogy in early nuclear theory or the roots of Yukawa's theory. *Revue d'Histoire des Sciences* 41: 225–97

Darrigol, O. 1988b. Elements of a scientific biography of Tomonaga Sin-itiro. *Historia Scientiarum* 35: 1–29

Debye, P. 1910. Der Wahrscheinlichkeitsbegriff in der Theorie der Strahlung. *Annalen der Physik* 23: 1427–34

Dirac, P. A. M. 1926a. On the theory of quantum mechanics. *Proceedings of the Royal Society of London (A)* 112: 661–77

Dirac, P. A. M. 1926b. The physical interpretation of the quantum mechanics. *Proceedings of the Royal Society of London (A)* 113: 621–41

Dirac, P. A. M. 1927a. The quantum theory of emission and dispersion. *Proceedings of the Royal Society of London (A)* 114: 243–65. Reprinted in Schwinger (1958)

Dirac, P. A. M. 1927b. The quantum theory of dispersion. *Proceedings of the Royal Society of London (A)* 114: 710–28

Dirac, P. A. M. 1928. The quantum theory of the electron. I. *Proceedings of the Royal Society of London (A)* 117: 610–24

Dirac, P. A. M. 1930a. A theory of electrons and protons. *Proceedings of the Royal Society of London (A)* 126: 360–5

Dirac, P. A. M. 1930b. On the annihilation of electrons and protons. *Proceedings of the Cambridge Philosophical Society* 26: 361–75

Dirac, P. A. M. 1930c. Note on the interpretation of the density matrix in the many-electron problem. *Proceedings of the Cambridge Philosophical Society* 27: 240–3

Dirac, P. A. M. 1931. Quantized singularities in the electromagnetic field. *Proceedings of the Royal Society of London (A)* 133: 60–72

Dirac, P. A. M. 1932. Relativistic quantum mechanics. *Proceedings of the Royal Society of London (A)* 136: 453–64

Dirac, P. A. M. 1933. Théorie du positron. In *Structure et Propriétés des Noyaux Atomiques: Rapports et Discussions du Septième Conseil de Physique tenù à Bruxelles du 22 au 29 octobre 1933*. Paris: Gauthier-Villars (Reprinted as [3], this volume)

Dirac, P. A. M. 1934. Discussion of the infinite distribution of electrons in the theory of the positron. *Proceedings of the Cambridge Philosophical Society* 30: 150–63. (Reprinted as [4], this volume)

Dirac, P. A. M., Fock, V. A. and Podolsky, B. 1932. On quantum electrodynamics. *Physikalische Zeitschrift der Sowjetunion* 2: 468–79. Reprinted in Schwinger (1958)

Dyson, F. J. 1949a. The radiation theories of Tomonaga, Schwinger and Feynman. *Physical Review* 75: 486–502. Reprinted in Schwinger (1958)

Dyson, F. J. 1949b. The *S* matrix in quantum electrodynamics. *Physical Review* 75: 1736–55 Reprinted in Schwinger (1958)

Dyson, F. J. 1951a. Heisenberg operators in quantum electrodynamics. I. *Physical Review* 82: 428–39

Dyson, F. J. 1951b. Heisenberg operators in quantum electrodynamics. II. *Physical Review* 83: 608–27

Ehrenfest, P. 1906. Zur Planckschen Strahlungstheorie. *Physikalische Zeitschrift* 7: 528–32

Ehrenfest, P. 1925. Energieschwankungen im Strahlungsfeld oder Kristallgitter bei Superposition quantisierter Eigenschwingungen. *Zeitschrift für Physik* 34: 362–73

Einstein, A. 1909. Entwicklung unserer Anschauungen über das Wesen und Konstitution der Strahlung. *Physikalische Zeitschrift* 10: 817–25

Einstein, A. 1925. Quantentheorie des idealen Gases. *Preussische Akademie der Wissenschaften, Phys.-math. Klasse, Sitzungsberichte*, pp. 18–25

Euler, H. 1936. Über die Streuung von Licht an Licht nach der Diracschen Theorie. *Annalen der Physik* 26: 398–448

Euler, H. and Kockel, B. 1935. Über die Streuung von Licht an Licht nach der Diracschen Theorie. *Die Naturwissenschaften* 15: 246–7

Evans, R. 1955. *The Atomic Nucleus*. New York: McGraw-Hill

Fermi, E. 1934. Versuch einer Theorie der β-Strahlen. I. *Zeitschrift für Physik* 88: 161–77

Feynman, R. P. 1949. The theory of positrons. *Physical Review* 76: 749–59

Feynman, R. P. 1965. The development of the space-time view of quantum electrodynamics. In *Nobel Lectures in Physics: 1963–1970* New York: North Holland

Galison, P. 1987. *How Experiments End*. University of Chicago Press

Gamow, G. 1931. *Constitution of Atomic Nuclei and Radioactivity*. Oxford: Clarendon

Gamow, G. 1937. *Structure of Atomic Nuclei and Nuclear Transformations*. Oxford: Clarendon

Gordon, W. 1926. Der Comptoneffekt nach der Schrödingerschen Theorie. *Zeitschrift für Physik* 40: 117–33

Harré, R. 1986. *Varieties of Realism*. London: Blackwell

Hartree, D. R. 1928. The wave mechanics of an atom with a non-Coulomb central field. Part I. Theory and methods. *Proceedings of the Cambridge Philosophical Society* 24: 89–110

Heisenberg, W. 1925. Über quantentheoretische Umdeutung kinematischer und mechanischer Beziehungen. *Zeitschrift für Physik* 33: 879–93 Translated in van der Waerden (1968)

Heisenberg, W. 1926a. Mehrkörperproblem und Resonanz in der Quantenmechanik. *Zeitschrift für Physik* 38: 411–26

Heisenberg, W. 1926b. Zur Quantenmechanik. *Die Naturwissenschaften* 14: 899–904

Heisenberg, W. 1926c. Schwankungserscheinungen und Quantenmechanik. *Zeitschrift für Physik* 40: 501–6

Heisenberg, W. 1927. Über den anschaulichen Inhalt der quantentheoretischen Kinematik und Mechanik. *Zeitschrift für Physik* 43: 172–98

Heisenberg, W. 1928. Zur Theorie des Ferromagnetismus. *Zeitschrift für Physik* 49: 619–36.

Heisenberg, W. 1929. Die Entwicklung der Quantentheorie, 1918–1928. *Die Naturwissenschaften* 26: 490–6

Heisenberg, W. 1930a. Die Selbstenergie des Elektrons. *Zeitschrift für Physik* 65: 4–13 (Reprinted as [1], this volume)

Heisenberg, W. 1930b. *Die physikalischen Prinzipien der Quantentheorie*. Leipzig: Herzl

Heisenberg, W. 1931. Bemerkungen zur Strahlungstheorie. *Annalen der Physik* 9: 338–46 (Reprinted as [2], this volume)

Heisenberg, W. 1932a. Über den Bau der Atomkerne. I. *Zeitschrift für Physik* 77: 1–11. Translated in Brink (1965), pp. 144–54

Heisenberg, W. 1932b. Über den Bau der Atomkerne. II. *Zeitschrift für Physik* 78: 156–64

Heisenberg, W. 1933a. Über den Bau der Atomkerne. III. *Zeitschrift für Physik* 80: 587–96

Heisenberg, W. 1933b. Considérations théorique générales sur la structure du noyau. In *Structure et Propriétés des Noyaux Atomiques: Rapports et Discussions du Septième Conseil de Physique tenù à Bruxelles du 22 au 29 octobre 1933*. Paris: Gauthier-Villars

Heisenberg, W. 1934a. Bemerkungen zur Diracschen Theorie des Positrons. *Zeitschrift für Physik* 98: 714–32 (Reprinted as [6], this volume)

Heisenberg, W. 1934b. Über die mit Enstehung von Materie aus Strahlung verknüpften Ladungsschwankungen. *Sachsische Akademie der Wissenschaften* 86: 317–22

Heisenberg, W. 1935. Bemerkungen zur Theorie des Atomkerns. In Pieter Zeeman: *Verhandelingen op 25 mei 1935 Aangeboden aan Prof. Dr. P. Zeeman*. The Hague: Nijhoff.

Heisenberg, W. 1936. Zur Theorie der 'Schauer' in der Höhenstrahlung. *Zeitschrift für Physik* 101: 533–40

Heisenberg, W. 1938. Über die in der Theorie der Elementarteilchen auftretende universelle Länge. *Annalen der Physik* 32: 20–33. (Reprinted as [10], this volume.)

Heisenberg, W. 1943a. Die 'beobachtbaren Grössen' in der Theorie der Elementarteilchen. *Zeitschrift für Physik* 120: 513–38

Heisenberg, W. 1943b. Die beobachtbaren Grössen in der Theorie der Elementarteilchen. *Zeitschrift für Physik* 120: 673–702

Heisenberg, W. 1946. Der mathematische Rahmen der Quantentheorie der Wellenfelder. *Zeitschrift für Naturforschung* 1: 608–22

Heisenberg, W. 1950. Zur Quantentheorie der Elementarteilchen. *Zeitschrift für Naturforschung* 5: 251–9

Heisenberg, W. and Euler, H. 1936. Folgerungen aus der Diracschen Theorie des Positrons. *Zeitschrift für Physik* 98: 714–32

Heisenberg, W. and Pauli, W. 1929. Zur Quantenelektrodynamik der Wellenfelder. *Zeitschrift für Physik* 56: 1–61

Heisenberg, W. and Pauli, W. 1930. Zur Quantentheorie der Wellenfelder II. *Zeitschrift für Physik* 59: 168–90

Heitler, W. 1928. Störungsenergie und austausch beim Mehrkörperproblem. *Zeitschrift für Physik* 46: 47–72

Heitler, W. 1936. *The Quantum Theory of Radiation*. Oxford: Clarendon

Heitler, W. 1954. *The Quantum Theory of Radiation*. New York: Dover

Heitler, W. and London, F. 1927. Wechselwirkung neutraler Atome und homöopolare

Bindung nach der Quantenmechanik. *Zeitschrift für Physik* 44: 455–72

Hendry, J. 1984. *The Creation of Quantum Mechanics and the Bohr–Pauli Dialogue*. Boston: Reidel

Holton, G. 1973. The roots of complementarity. In *Thematic Origins of Scientific Thought: Kepler to Einstein*. Cambridge, MA: Harvard University Press

Hovis, R. C. and Kragh, H. 1991. Resource Letter HPP-1: History of elementary-particle physics. *American Journal of Physics* 59: 779–807

Iwanenko, D. 1932. The neutron hypothesis. *Nature* 129: 798

Iwanenko, D. 1934. Interaction of neutrons and protons. *Nature* 133: 981–2

Jammer, M. 1966. *The Conceptual Development of Quantum Mechanics*. New York: McGraw-Hill

Jammer, M. 1974. *The Philosophy of Quantum Mechanics*. New York: Wiley

Jensen, H. J. 1963. Glimpses at the history of nuclear theory. In *Nobel Lectures in Physics*. New York: North-Holland

Jordan, P. 1936. *Anschauliche Quantentheorie: eine Einführung in die moderne Auffassung der Quantenerscheinungen*. Berlin: Springer

Jordan, P. and Klein, O. 1927. Zum Mehrkörperproblem der Quantentheorie. *Zeitschrift für Physik* 45: 755–65

Jordan, P. and Pauli, W. 1928. Zur Quantenelektrodynamik ladungsfreier Felder. *Zeitschrift für Physik* 47: 151–73

Jordan, P. and Wigner, E. 1928. Über das Paulische Äquivalenzverbot. *Zeitschrift für Physik* 47: 631–51. Reprinted in Schwinger (1958).

Kemmer, N. and Weisskopf V. F. 1936. Deviations from the Maxwell equations resulting from the theory of the positron. *Nature* 137: 659

Klein, O. 1926. Quantentheorie und fünfdimensionale Relativitätstheorie. *Zeitschrift für Physik* 37: 895–906

Klein, O. 1928. Die Relexion von Elektronen an einem Potentialsprung nach der relativistischen Dynamik von Dirac. *Zeitschrift für Physik* 53: 157–65

Klein, O. and Nishina, Y. 1928. Über die Streuung von Strahlung durch freie Elektronen nach der neuen relativistischen Quantenelektrodynamik von Dirac. *Zeitschrift für Physik* 52: 853–68

Kragh, H. 1981. The genesis of Dirac's theory of electrons. *Archive for History of Exact Sciences* 24: 31–67

Kragh, H. 1989. The negative proton: its earliest history. *American Journal of Physics* 57: 1034–9

Kramers, H. 1924. The quantum theory of dispersion. *Nature* 1 14: 310–11

Kramers, H. and Heisenberg, W. 1925. Über die Streuung von Strahlung durch Atome. *Zeitschrift für Physik* 31: 681–707. Translated in van der Waerden (1968)

Ladenburg, R. 1921. Die quantentheoretische Deutung der Zahl der Dispersionelektronen. *Zeitschrift für Physik* 4: 451–68. Reprinted in van der Waerden (1968)

Ladenburg, R. and Reiche, F. 1923. Absorption, Zerstreuung und Dispersion in der Bohrschen Atomtheorie. *Die Naturwissenschaften* 11: 584–98

Landau, L. and Peierls, R. 1930. Quantenelektrodynamik in Konfigurationraum. *Zeitschrift für Physik* 62: 188–200

Landau, L. and Peierls, R. 1931. Erweiterung des Unbestimmtheitsprinzips für die relativistische Quantentheorie. *Zeitschrift für Physik* 69: 56–69

Lorentz, H. A. 1904. Electromagnetic phenomena in a system moving with any velocity less than that of light. *Proceedings of the Royal Academy of Amsterdam* 6: 809–34

Majorana, E. 1933. Über die Kerntheorie. *Zeitschrift für Physik* 82: 137–45. Translated in Brink (1965)

with

$$\left.\begin{aligned} U_F(r, t) &= U(r, t) - U(r, -t) \\ &= 2J_0\{m(t^2 - r^2)^{1/2}/\hbar\} && t > r \\ &= 0 && r > t > -r \\ &= -2J_0\{m(t^2 - r^2)^{1/2}/\hbar\} && -r > t \end{aligned}\right\}. \tag{19}$$

Similarly the distribution R_1 will be given by

$$(x't'|R_1|x''t'') = -\left[i\hbar\frac{\partial}{\partial t} - i\hbar\alpha_s\frac{\partial}{\partial x_s} + \alpha_4 m\right]S_1(x, t), \tag{20}$$

$$S_1(x, t) = \frac{-i}{4h}\frac{1}{r}\frac{\partial}{\partial r}U_1(r, t), \tag{21}$$

with

$$\left.\begin{aligned} U_1(r, t) &= U(r, t) + U(r, -t) \\ &= 2iY_0\{m(t^2 - r^2)^{1/2}/\hbar\} && t > r \\ &= 2H_0^{(1)}\{im(r^2 - t^2)^{1/2}/\hbar\} && r > t > -r \\ &= 2iY_0\{m(t^2 - r^2)^{1/2}/\hbar\} && -r > t \end{aligned}\right\}. \tag{22}$$

It is clear from these equations that there will be singularities, not only at the point $x_s = 0$, $t = 0$, but also everywhere on the light-cone $t^2 - r^2 = 0$. In order to determine these singularities, we may expand the Bessel functions in power series of $(t^2 - r^2)^{1/2}$ and retain only the first few terms. If we retain only the first term in (19), we get for U_F the constant value zero outside the light-cone and the constant value 2, -2 in the two regions inside the light-cone for which $t > r$ and $t < -r$, respectively. This U_F substituted in (18) gives

$$S_F(x, t) = (i/2hr)\{\delta(r - t) - \delta(r + t)\}.$$

For $t > 0$, this may be written

$$S_F(x, t) = (i/2hr)\delta(r - t) = (i/h)\delta(t^2 - r^2),$$

and when substituted in (17) gives

$$(x't'|R_F|x''t'') = (1/\pi)(t + \alpha_s x_s)\delta'(t^2 - r^2), \tag{23}$$

with neglect of a term involving $\delta(t^2 - r^2)$. This is the worst singularity of R_F. For $t < 0$ the result is the same, except for a change of sign. If we retained the second term in the expansion of J_0 in (19), we should get the next worst singularity of R_F, containing $\delta(t^2 - r^2)$. If we retained also the third term in J_0 in (19), we should get also the third worst kind of singularity in R_F, involving a plain discontinuity on the light-cone. The fourth and higher terms in J_0 would not give rise to any singularity in R_F. In this way we can determine completely all the singularities in R_F.

Let us now examine the singularities in R_1. The important terms in U_1, given by (22), are

$$(4i/\pi)\log\{m(t^2 - r^2)^{1/2}/\hbar\}J_0\{m(t^2 - r^2)^{1/2}/\hbar\} \qquad |t| > r, \tag{24}$$

$$[2 + (4i/\pi)\log\{im(r^2 - t^2)^{1/2}/\hbar\}]J_0\{im(r^2 - t^2)^{1/2}/\hbar\} \qquad |t| < r. \tag{25}$$

The remaining terms are a power series in $t^2 - r^2$, of the same form inside and outside the light-cone, and therefore they do not give rise to any singularity. Expression (25) may be simplified to

$$(4i/\pi)\log\{m(r^2 - t^2)^{1/2}/\hbar\}J_0\{im(r^2 - t^2)^{1/2}/\hbar\} \qquad |t| < r.$$

Substituting this and (24) into (21) we get, if we take only the first term of J_0 into account,

$$S_1(x, t) = \frac{1}{\pi h}\frac{1}{r^2 - t^2}. \tag{26}$$

It is important to see that no δ function occurs in $S_1(x, t)$. The reason at the bottom of this is that, in differentiating the logarithm that occurs in the Bessel functions Y_0 and $H_0^{(1)}$, we must use the formula

$$\frac{d}{dz}\log z = \frac{1}{z} - i\pi\delta(z) \tag{27}$$

in which the term $-i\pi\delta(z)$ is required in order to make the integral of the right-hand side of (27) between the limits a and $-a$ equal $\log(-1)$, the integral of $1/z$ between these limits being assumed to be zero. The δ function which arises in this way just cancels with that arising from the fact that we have $H_0^{(1)}$ instead of Y_0 in (22) when $r > t > -r$. This cancellation is exact and still holds when (22) is differentiated more than once with respect to any of the variables t, r, x_s.

On substituting (26) into (20), we find

$$(x't'|R_1|x''t'') = \frac{-i}{\pi^2}\frac{t + \alpha_s x_s}{(t^2 - r^2)^2}, \tag{28}$$

neglecting a term involving $1/(t^2 - r^2)$. This gives the worst singularity of R_1. By taking into account the second and third terms in the expansion of the J_0 in (24) and (25), we can calculate the other singularities in R_1, involving $1/(t^2 - r^2)$ and $\log|t^2 - r^2|$.

The main result of this investigation for the case of no field is that there are two quite distinct kinds of singularity occurring in the matrices R_F and R_1 respectively. The singularities occurring in R_F are all associated with the δ function and those in R_1 with the reciprocal function and logarithm. From the generality of this result we may expect it to hold also when there is a field present.

3 Case of an arbitrary field

Let us now examine the singularities in $(x't'|R_F|x''t'')$ and $(x't'|R_1|x''t'')$ when there is a general field present. Our method will be to suppose that the singularities are of the same form as in the case of no field, but have unknown coefficients. These coefficients must be functions of x'_s, t', x''_s, t'' which are free from singularities and can be expanded as Taylor series for small values of x_s and t. We must try to choose them so that the equations of motion (8) are satisfied.

The application of the method follows a parallel course for R_F and R_1, and we need therefore treat in detail only R_1, which is the density matrix we are mainly interested in. We put

$$(x't'|R_1|x''t'') = u\,\frac{t + \alpha_s x_s}{(t^2 - r^2)^2} + \frac{v}{t^2 - r^2} + w\log|t^2 - r^2|, \tag{29}$$

where u, v and w are functions of x'_s, t', x''_s, t'' or of x_s, t, x''_s, t'', which are free from singularities for small values of x_s and t. To get sufficient generality we must allow u, v and w to be matrices in the spin variables and thus not necessarily to commute with the α's. We shall find, however, that we can satisfy all the conditions with u diagonal in the spin variables and thus commuting with all the α's. For the sake of brevity in the algebraic work, we shall assume already now that u is diagonal in the spin variables.

We must now try to choose u, v and w so that the second of equations (8) is satisfied. It is convenient to use the symbol $\mathcal{H}$ to denote the differential operator

$$\mathcal{H} = \mathrm{i}\hbar\left(\frac{\partial}{\partial t} + \alpha_s \frac{\partial}{\partial x_s}\right) + e(A_0 - \alpha_s A_s) - \alpha_4 m, \tag{30}$$

the function that it operates on being assumed to be expressed in terms of the variables x_s, t, x''_s, t'' and not to involve x'_s, t' explicitly. With this notation, we get from (8)

$$\mathcal{H}\left\{u\frac{t + \alpha_s x_s}{(t^2 - r^2)^2} + \frac{v}{t^2 - r^2} + w\log|t^2 - r^2|\right\} = 0,$$

which reduces to

$$(\mathcal{H}u)\frac{t + \alpha_s x_s}{(t^2 - r^2)^2} + (\mathcal{H}v)\frac{1}{t^2 - r^2} - 2\mathrm{i}\hbar\frac{t - \alpha_s x_s}{(t^2 - r^2)^2}v$$
$$+ (\mathcal{H}w)\log|t^2 - r^2| + 2\mathrm{i}\hbar\frac{t - \alpha_s x_s}{t^2 - r^2}w = 0. \tag{31}$$

In order that this equation may hold when u, v and w are free from singularity, the log term must vanish by itself and hence

$$\mathcal{H}w = 0 \tag{32}$$

and

$$(\mathcal{H}u)(t + \alpha_s x) + (\mathcal{H}v)(t^2 - r^2) - 2\mathrm{i}\hbar(t - \alpha_s x)v$$
$$+ 2\mathrm{i}\hbar(t - \alpha_s x_s)w(t^2 - r^2) = 0. \tag{33}$$

Now from the commutability relations for the α's we find, after a simple calculation,

$$(\mathcal{H}u)(t + \alpha_x x_s) = -(t - \alpha_s x_s)\mathcal{G}u$$
$$+ 2\left\{t\left(\mathrm{i}\hbar\frac{\partial}{\partial t} + eA_0\right) + x_s\left(\mathrm{i}\hbar\frac{\partial}{\partial x_s} - eA_s\right)\right\}u, \tag{34}$$

where $\mathcal{G}$ denotes the differential operator

$$\mathcal{G} = \mathrm{i}\hbar\left(\frac{\partial}{\partial t} - \alpha_s\frac{\partial}{\partial x_s}\right) + e(A_0 + \alpha_s A_s) + \alpha_4 m. \tag{35}$$

If we substitute the right-hand side of (34) for the first term in (33), we get an equation which may be written

$$2\left\{t\left(\mathrm{i}\hbar\frac{\partial}{\partial t} + eA_0\right) + x_s\left(\mathrm{i}\hbar\frac{\partial}{\partial x_s} - eA_s\right)\right\}u + (t - \alpha_s x_s)B = 0, \tag{36}$$

where

$$B = (t + \alpha_s x_s)\mathcal{H}v - 2\mathrm{i}\hbar v + 2\mathrm{i}\hbar w(t^2 - r^2) - \mathcal{G}u. \tag{37}$$

When a given $(x't'|R_1|x''t'')$ is expressed in the form (29), u and v are not completely determined, since we can always add a term of the form $b(t - \alpha_s x_s)$ to u and subtract b from v without changing the right-hand side of (29), b being anything free from singularities. It can easily be shown that we may choose b so as to make $B = 0$. With this extra condition, equation (36) can be solved for u, the result being

$$u = k \exp\left\{ie \int (A_0 \,\mathrm{d}t - A_s \,\mathrm{d}x_s)/\hbar\right\}, \tag{38}$$

where k is an arbitrary coefficient and the integral is taken along the straight line in space-time joining the point x_s'', t'' to the point x_s', t'. We must take k equal to $-\mathrm{i}/\pi^2$ so as to make the worst singularity of the right-hand side of (29) equal to the right-hand side of (28) for small values of x_s, t. This determines u completely.

We must now deal with equation (37) with $B = 0$. It may be written

$$(t + \alpha_s x_s) f - 2\mathrm{i}\hbar v - \mathcal{G} u = 0, \tag{39}$$

where

$$f = \mathcal{H} v + 2\mathrm{i}\hbar(t - \alpha_s x_s) w. \tag{40}$$

Equation (39) gives v in terms of f and allows us to eliminate v and work with the unknown f instead. Eliminating v from (40), we get

$$2\mathrm{i}\hbar f = \mathcal{H}(t + \alpha_s x_s) f - \mathcal{H}\mathcal{G} u - 4\hbar^2 (t - \alpha_s x_s) w. \tag{41}$$

Now by the same kind of calculation as led to (34) we find

$$\begin{aligned}\mathcal{H}(t + \alpha_s x_s) f = {} & -(t - \alpha_s x_s)\mathcal{G} f + 4\mathrm{i}\hbar f \\ & + 2\left\{t\left(\mathrm{i}\hbar \frac{\partial}{\partial t} + eA_0\right) + x_s\left(\mathrm{i}\hbar \frac{\partial}{\partial x_s} - eA_s\right)\right\} f.\end{aligned} \tag{42}$$

If this expression is substituted for the first term in the right-hand side of (41), we get the result

$$\begin{aligned}& 2\left\{t\left(\mathrm{i}\hbar \frac{\partial}{\partial t} + eA_0\right) + x_s\left(\mathrm{i}\hbar \frac{\partial}{\partial x_s} - eA_s\right)\right\} f + 2\mathrm{i}\hbar f \\ & \quad = \mathcal{H}\mathcal{G} u + (t - \alpha_s x_s)(4\hbar^2 w + \mathcal{G} f).\end{aligned} \tag{43}$$

A way of solving this equation for f is first to solve the corresponding equation with the term containing the factor $t - \alpha_s x_s$ omitted, i.e. the equation

$$2\left\{t\left(\mathrm{i}\hbar \frac{\partial}{\partial t} + eA_0\right) + x_s\left(\mathrm{i}\hbar \frac{\partial}{\partial x_s} - eA_s\right)\right\} f_1 + 2\mathrm{i}\hbar f_1 = \mathcal{H}\mathcal{G} u. \tag{44}$$

Then f will be of the form

$$f = f_1 + (t - \alpha_s x_s) g, \tag{45}$$

where g is free from singularities. Substituting (45) in (43) and using (44), we get, after cancelling the factor $t - \alpha_s x_s$,

$$\begin{aligned}& 2\left\{t\left(\mathrm{i}\hbar \frac{\partial}{\partial t} + eA_0\right) + x_s\left(\mathrm{i}\hbar \frac{\partial}{\partial x_s} - eA_s\right)\right\} g + 4\mathrm{i}\hbar g \\ & \quad = 4\hbar^2 w + \mathcal{G} f_1 + \mathcal{G}(t - \alpha_s x_s) g.\end{aligned}$$

If we now use an equation like (42), with g instead of f and the signs of all the α's changed, this reduces to

$$(t + \alpha_s x_s)\mathcal{H} g = 4\hbar^2 w + \mathcal{G} f_1. \tag{46}$$

Equation (44) fixes f_1 completely, without any arbitrary constant. One can see this by substituting for the various functions in this equation their Taylor expansions in powers of x_s and t, when it will be found that the coefficients of f_1 of the nth degree are completely determined in terms of those of lower degree and of the coefficients of the A's.

We are thus left with the problem of solving equations (32) and (46) for w and g. If we apply the operator $\mathcal{H}$ to equation (46), the term involving w drops out by equation (32), leaving

$$\mathcal{H}(t + \alpha_s \dot{x}_s)\mathcal{H}g = \mathcal{H}\mathcal{G}f_1. \tag{47}$$

This, considered as an equation in the unknown $\mathcal{H}g$, determines $\mathcal{H}g$ completely, as may also be verified by Taylor expansions. From (46), w is now determined. In this way all our equations are satisfied and all our unknowns are determined, with the exception of g, which is not itself determined although $\mathcal{H}g$ is. The final result is

$$(x't'|R_1|x''t'') = u\,\frac{t + \alpha_s x_s}{(t^2 - r^2)^2} + \frac{(t + \alpha_s x_s)f_1 - \mathcal{G}u}{2i\hbar(t^2 - r^2)} + \frac{g}{2i\hbar} + w\log|t^2 - r^2|, \tag{48}$$

where u is given by (38), and f_1, $\mathcal{H}g$ and w are determined by (44), (47) and (46) respectively.

To do the corresponding work for R_{F} we put, analogously to (29),

$$(x't'|R_{\mathrm{F}}|x''t'') = u(t + \alpha_s x_s)\delta'(t^2 - r^2) + v\delta(t^2 - r^2) + w\gamma(t^2 - r^2), \tag{49}$$

where γ is the function

$$\gamma(z) = 0 \qquad z < 0$$
$$\gamma(z) = 1 \qquad z > 0,$$

and we again try to choose u, v and w to satisfy the equation of motion (8). The equations that we now get for u, v and w are exactly the same as before and thus their solution will be the same, or else will differ from the previous solution by a numerical factor. In order that the worst singularity of the right-hand side of (49), i.e. the first term in it, may be the same as the right-hand side of (23) for small values of x_s, t, we must choose this numerical factor equal to $i\pi$. Hence expressison (49) with u, v and w equal to $i\pi$ times their values in (29) must give the matrix elements $(x't'|R_{\mathrm{F}}|x''t'')$ with $t > 0$. The elements with $t < 0$ are given by (49) with u, v and w equal to $-i\pi$ times their values in (29).

The indeterminacy which we have in g will not lead to any indeterminacy in $(x't'|R_{\mathrm{F}}|x''t'')$, since a variation in g causes v to change by a term containing $t^2 - r^2$ as a factor and such a term multiplied into $\delta(t^2 - r^2)$ will vanish. Thus R_{F} is completely fixed.

4 Conclusion

From the foregoing work we see that the following results must hold, at least to the accuracy of the Hartree method of approximation:

(i) One can give a precise meaning to the distribution of electrons in which every state is occupied. This distribution may be defined as that described by the density matrix R_F given by (49), this matrix being completely fixed for any given field.

(ii) One can give a precise meaning to a distribution of electrons in which nearly all (i.e. all but a finite number, or all but a finite number per unit volume) of the negative-energy states are occupied and nearly all of the positive-energy ones unoccupied. Such a distribution may be defined as one described by a density matrix $R = \frac{1}{2}(R_F + R_1)$, where R_1 is of the form (48). This definition is permissible because the only possible variations in R_1, namely those due to g not being completely defined, are free from singularity and thus correspond to finite changes, or finite changes per unit volume, in the electron distribution. Our method does not give any precise meaning to which negative-energy states are unoccupied or which positive-energy ones are occupied. It is sufficiently definite, though, to take as the basis of the theory of the positron the assumption that only distributions described by $R = \frac{1}{2}(R_F + R_1)$ with R_1 of the form (48) occur in nature.

(iii) A distribution R such as occurs in nature according to the above assumption can be divided naturally into two parts

$$R = R_a + R_b,$$

where R_a contains all the singularities and is also completely fixed for any given field, so that any alteration one may make in the distribution of electrons and positrons will correspond to an alteration in R_b but to none in R_a. We get this division into two parts by putting the term containing g into R_b and all the other terms into R_a. Thus

$$R_b = g/4i\hbar.$$

It is easily seen that R_b is relativistically invariant and gauge invariant, and it may be verified after some calculation that R_b is Hermitian and that the electric density and current density corresponding to it satisfy the conservation law (9). It therefore appears reasonable to make the assumption that *the electric and current densities corresponding to R_b are those which are physically present, arising from the distribution of electrons and positrons*. In this way we can remove the infinities mentioned at the end of §1.

The present paper is incomplete in that the effect of the exclusion principle, equation (2) or (6), on R_b has not been investigated. Further work that remains to be done is to examine the physical consequences of the foregoing assumption and to see whether it leads to any phenomena of the nature of a polarization of a vacuum by an electromagnetic field.

5 The self-energy of the electron

V. WEISSKOPF

Zeitschrift für Physik, 89: 27–39 (1934). Received 13 March 1934.

The self-energy of the electron is derived in a closer formal connection with classical radiation theory, and the self-energy of an electron is calculated when the negative energy states are occupied, corresponding to the conception of positive and negative electrons in the Dirac 'hole' theory. As expected, the self-energy also diverges in this theory, and specifically to the same extent as in ordinary single-electron theory.

1 Problem definition

The self-energy of the electron is the energy of the electromagnetic field which is generated by the electron in addition to the energy of the interaction of the electron with this field. Waller,[1] Oppenheimer,[2] and Rosenfeld[3] calculated the self-energy of the free electron by means of the Dirac relativistic wave equation of the electron and the Dirac theory of the interaction between matter and light. They here used an approximation method which represents the self-energy in powers of the charge e. They found that the first term, which is proportional to e^2, already becomes infinitely large. The essential reason for this is that the theory of the interaction of the electron with the electromagnetic field is built on the classical equations of motion of a point-shaped electron whose self-energy, as is well known, also becomes infinite in classical theory.[4]

In the present note, the expressions for the self-energy shall be derived without direct application of quantum electrodynamics, but by means of the Heisenberg radiation theory,[5] which is linked much more closely to classical electrodynamics. The radiation field is calculated classically from the current and charge densities of the atom; however, the amplitudes of the electromagnetic potentials are regarded as non-commuting in the final result. Just as was shown in a corresponding paper by Casimir[6] concerning the natural linewidth, this method yields the same result as explicit quantum

[1] I. Waller, *ZS. f. Phys.* **62**, 673, 1930.

[2] R. Oppenheimer, *Phys. Rev.* **35**, 461, 1930.

[3] L. Rosenfeld, *ZS. f. Phys.* **70**, 454, 1931.

[4] Recently, G. Wentzel (*ZS. f. Phys.* **86**, 479, 635, 1933) has shown that one can circumvent the divergence of the self-energy in classical electron theory by suitable limiting processes. The transfer of these methods to quantum theory has failed, however, since, according to Waller, the degree of infinity in quantum theory is higher than in classical theory. The hope expressed there that the degree of infinity will become smaller in the Dirac formalism of the 'hole' theory, does indeed hold for the electrostatic part but not for the electrodynamic part, so that the Wentzel method must fail here too.

[5] W. Heisenberg, *Ann. d. Phys.* **338**, 1931; see also W. Pauli's article in Geiger-Scheel, *Handb. d. Phys.* XXIC/1, 2nd edn., pp. 201–10.

[6] H. Casimir, *ZS. f. Phys.* **81**, 496, 1933.

electrodynamics and thus leads to the same difficulties. It is equivalent to the latter in every respect. It therefore perhaps is not quite appropriate to designate this method as a correspondence method in contrast to the Dirac radiation theory (see Casimir, *loc. cit.*), if, with Bohr, one sees the essential content of the correspondence argument in the tendency to preserve the consequences of classical theory as long as they do not directly contradict the specific atom-mechanical phenomena.

Furthermore, the calculation shall be performed generally for a multi-electron system; in particular, it will also be evaluated for the case that all negative energy states are occupied. The self-energy of the electron in the new Dirac 'hole' theory can then be calculated. As is well known, this theory assumes – in order to eliminate the difficulties of negative energy states – that these states are generally occupied. However, only a certain residue of the density-current vector, which comes from all the existing electrons, becomes noticeable. This residue is obtained by subtracting that density-current vector which is generated by the complete occupation of all negative states of the electron. Consequently, if a negative energy state has remained unoccupied, for example, one obtains a residual amount which corresponds to the current-density vector of a positive electron.

In calculating the self-energy, we shall limit ourselves to terms proportional to e^2 in the expansion by powers of the charge. Consequently, the calculation also does not contain those difficulties and ambiguities which appear in the action of electromagnetic fields upon electrons in negative states, since, in the required approximation, these may be assumed to be empty.

2 The self-energy of the electron

The energy operator of an electron with a surrounding field is

$$H = c\left(\vec{\alpha}, \vec{p} - \frac{e}{c}\vec{A}\right) + \beta mc^2 + \frac{1}{8\pi}\int(\vec{E}^2 + \vec{H}^2)\,\mathrm{d}\vec{r},$$

whereby $\vec{\alpha}$ and β are the familiar Dirac spin matrices, $\vec{A}$ is the vector potential of the fields $\vec{E}, \vec{H}$. It should be emphasized that the scalar potential Φ does not occur at all in this expression.[7] The self-energy is then given by the expectation value of

$$\mathrm{E} = -\int(\vec{i}\vec{A})\,\mathrm{d}\vec{r} + \frac{1}{8\pi}\int(\vec{E}^2 + \vec{H}^2)\,\mathrm{d}\vec{r}$$

where $\vec{i} = e\vec{\alpha}$ is the current-density operator and $\vec{A}, \vec{E}, \vec{H}$ here describe the field which is generated by the current-density $\boldsymbol{i}$ and the charge density ρ.

We now divide the electromagnetic field into a rotation-free part with components $\vec{A}_1$, $\vec{E}_1$ ($\vec{H}_1 = 0$) and a divergence-free part[8] $\vec{A}_{\mathrm{tr}}$, $\vec{E}_{\mathrm{tr}}$, $\vec{H}_{\mathrm{tr}}$. The 'electrostatic' self-energy coming from the first part

$$E^{\mathrm{S}} = -\int(\vec{i}\vec{A}_1)\,\mathrm{d}\vec{r} + \frac{1}{8\pi}\int\vec{E}_1^2\,\mathrm{d}\vec{r}$$

[7] $\mathrm{d}\vec{r}$ signifies the volume element $\mathrm{d}x\,\mathrm{d}y\,\mathrm{d}z$.

[8] The indices come from the circumstance that, in a spatial Fourier decomposition, the former are represented by transverse [tr] waves, whereas the latter are represented by longitudinal waves.

can easily be brought into the following form:

$$E^{\mathrm{S}} = \frac{1}{2}\int \rho\Phi' \, \mathrm{d}\vec{r}, \tag{1}$$

by regauging the potentials, by means of the expressions $\vec{A}_{\mathrm{l}} = \operatorname{grad}\lambda$, $\Phi' = \Phi - \lambda$, (and thus $\vec{E}_{\mathrm{l}} = -\operatorname{grad}\varphi'$). Here,

$$\Phi'(\vec{r}) = \int \frac{\rho(\vec{r}')}{|\vec{r} - \vec{r}|} \mathrm{d}\vec{r}'. \tag{1a}$$

The 'electrodynamic' self-energy which comes from the divergence-free part of the field

$$E^{\mathrm{P}} = -\int (\vec{i}_{\mathrm{tr}}\vec{A}_{\mathrm{tr}}) \, \mathrm{d}\vec{r} + \frac{1}{8\pi}\int (E_{\mathrm{tr}}^2 + H_{\mathrm{tr}}^2) \, \mathrm{d}\vec{r}$$

can be reduced to the time average

$$E^{\mathrm{D}} = -\frac{1}{2}\int (\vec{i}_{\mathrm{tr}}\vec{A}_{\mathrm{tr}}) \, \mathrm{d}\vec{r} \tag{2}$$

by means of the following consideration, in the approximation which is solely being considered here. (Here, $\vec{i}_{\mathrm{tr}}$ is the divergence-free part of the current density). It should be noted that $\vec{A}_{\mathrm{tr}}$ is a gauge-invariant quantity.

According to a well-known theorem from analytical mechanics, the change in energy through the adiabatic change of a parameter a is given by

$$\Delta E = \overline{\frac{\partial H(p, q, a)}{\partial a}} \Delta a,$$

where the bar denotes time average or, in quantum theory, the diagonal element. We will regard $\boldsymbol{i}_{\mathrm{tr}}$ as a parameter independent of the remaining variables, since only $\boldsymbol{i}_{\mathrm{l}}$ is determined by the continuity equation. If we set $\boldsymbol{i}_{\mathrm{tr}} = \lambda \boldsymbol{i}_{\mathrm{tr}}^0$, and if we allow the current $\boldsymbol{i}_{\mathrm{tr}}$ to increase from zero to its actual value $\boldsymbol{i}_{\mathrm{tr}}^0$ through a change in λ, and if we use as a basis the expansion

$$H' = -\int \vec{i}_{\mathrm{tr}}\vec{A}_{\mathrm{tr}} \, \mathrm{d}\vec{r} = \lambda H_1' + \lambda^2 H_2' + \cdots,$$

where H_0' is proportional to the charge e, and H_1' is proportional to e^2, we obtain the expression:

$$E^{\mathrm{D}} = \int_0^1 \overline{\frac{\partial H}{\partial \boldsymbol{i}_{\mathrm{tr}}}} \boldsymbol{i}_{\mathrm{tr}}^0 \, \mathrm{d}\lambda = \overline{H}_1' + \frac{1}{2}\overline{H}_2' + \cdots.$$

The first term is proportional to the charge e and must vanish because of the symmetry of the self-energy with respect to the sign of e; if we limit ourselves to terms that are quadratic in e^2, we thus obtain $E^{\mathrm{D}} = \frac{1}{2}\overline{H}'$.

In the hole theory of the electron, we must write instead of (1) and (2):

$$E^{\mathrm{S}} = \frac{1}{2}\int\int \frac{[\rho(\vec{r}) - \tilde{\rho}(\vec{r})][\rho(\vec{r}') - \tilde{\rho}(\vec{r}')]}{|\vec{r} - \vec{r}'|} \mathrm{d}\vec{r} \, \mathrm{d}\vec{r}', \tag{3}$$

$$E^{\mathrm{D}} = -\frac{1}{2}\int (\vec{\imath}(\vec{r}), - \vec{\tilde{\imath}}(\vec{r}), \vec{A}(\vec{r}) - \vec{\tilde{A}}(\vec{r})) \, \mathrm{d}\vec{r}, \tag{4}$$

where $\tilde{\rho}$ and $\vec{\tilde{i}}$ denote those charge and current densities of electrons in negative states which must be subtracted off and where $\vec{\tilde{A}}$ denotes the vector potential that is generated. Furthermore, we must note that, in this theory, a self-energy is also associated with the 'vacuum', i.e. that state in which all negative energy levels are occupied. As the self-energy of an electron we must then take the difference between the state with a normal electron of positive energy when the negative energy levels are occupied and the self-energy of the 'vacuum'.

3 Calculation of the self-energy

In calculating the energy expressions (3) and (4), we start from the stationary states of a free electron, which result from the Dirac equation

$$\frac{\hbar c}{\mathrm{i}}(\vec{\alpha}, \mathrm{grad})\varphi + \beta mc^2\varphi = E\varphi$$

The eigenfunctions are[9]

$$\varphi^k(\vec{p}\vec{r}) = \frac{1}{h^3}u^k(\vec{p})\exp\left(\frac{\mathrm{i}}{\hbar}(\vec{p}\vec{r})\right), \; k = 1, 2, 3, 4, \tag{5}$$

with the eigenvalues

$$E^k(p) = +\,c(p^2 + m^2c^2)^{1/2} \text{ for } k = 1, 2,$$
$$E^k(p) = -\,c(p^2 + m^2c^2)^{1/2} \text{ for } k = 3, 4.$$

The functions φ^k and the quantities $u^k(\vec{p})$ are four-component quantities and can be represented as vectors in four-dimensional spin space. The scalar product of two such quantities is written in the form[10]

$$\{u^{k*}(p), u^l(p')\} = \sum_{\rho=1}^{4} u_\rho^{k*}(p)u_\rho^l(p'),$$

where $u_\rho^k(p)$, $\rho = 1, \ldots, 4$, are the components of $u^k(p)$. The following always holds:

$$\{u^{k*}(\vec{p})u^l(\vec{p})\} = \delta_{kl}, \; \sum_{k=1}^{4} u_\rho^{k*}(\vec{p})u_\sigma^k(\vec{p}) = \delta_{\rho\sigma}. \tag{6}$$

With these relations, the normalization of the eigenfunctions is fixed in the usual fashion so that

$$\int\{\varphi^{k*}(\vec{p}\vec{r}), \varphi^l(\vec{p}'\vec{r})\}\,\mathrm{d}\vec{r} = \delta(\vec{p} - \vec{p}')\delta_{kl}. \tag{6'}$$

A normalized eigenfunction then just represents an electron in the volume h^3. The values of the self-energy obtained with this normalization thus must be multiplied by h^3/V, where V is the total volume, so that these values can refer to a single electron.

To calculate the self-energy of a multi-electron system, it is advantageous to use the method of quantized waves, in which the charge and current densities act as operators on the eigenfunctions, whose variables are the occupation numbers $N^k(\vec{p})$ of the stationary states $\vec{p}^k$, $k = 1, \ldots, 4$ of the free electron. The following then holds:

$$\left.\begin{aligned} \rho(\vec{r}) &= \{\psi^\dagger(\vec{r}), \psi(\vec{r})\}, \\ \vec{i}(\vec{r}) &= \{\psi^\dagger(\vec{r}), \alpha\psi(\vec{r})\} \end{aligned}\right\} \tag{7}$$

[9] $2\pi\hbar = h$ = Planck's constant. Below we shall use both symbols $\hbar$ and h.

[10] Where no misunderstanding is possible, the arrow above the letter is omitted.

with

$$\psi = \sum_k \int d\vec{p}\, a^k(\vec{p})\varphi^k(\vec{p}\vec{r}). \tag{8}$$

The $a^k(\vec{p})$ are operators which act on the eigenfunctions $c(\ldots N^k(p)\ldots)$ in the following fashion:

$$a^k(p)\cdot c(\cdots N^k(p)\cdots) = \varepsilon N^k(p)c(\cdots 1 - N^k(p)\cdots),$$
$$a^{\dagger k}(p)\cdot c(\cdots N^k(p)\cdots) = \varepsilon(1 - N^k(p))\cdot c(\cdots 1 - N^k(p)\cdots),$$

where ε can have the value $+1$ or -1. We therefore have

$$\left.\begin{aligned} a^{k\dagger}(p)a^k(p) &= N^k(p),\\ a^k(p)a^{k\dagger}(p) &= 1 - N^k(p). \end{aligned}\right\} \tag{9}$$

(a) The electrostatic self-energy

We first calculate the electrostatic portion of the self-energy E^{S}, which is associated with a particular occupation of the states. To save indices, we shall until further notice designate the states $\vec{\mathrm{p}}^k$ with the letters $q, r, s \ldots$, and by $\overset{+}{q}, \overset{+}{r}$, and respectively $\bar{q}, \bar{r}$, we shall understand states of positive or negative energy. The sum over k in the integral of p in (8) shall be replaced by a single symbolic sum:

$$\psi(\vec{r}) = \sum_q a_q\varphi_q(\vec{r}).$$

The density $\tilde{\rho}$ of the electrons in negative energy states then is

$$\tilde{\rho}(\vec{r}) = \sum_{\bar{q}}\{\varphi^*_{\bar{q}}(\vec{r}), \varphi_{\bar{q}}(\vec{r})\}. \tag{10}$$

By using the quantity

$$A_{qrst} = e^2 \int \frac{\{\varphi^*_q(\vec{r}), \varphi_r(\vec{r})\}\{\varphi^*_s(\vec{r}\,'), \varphi_t(\vec{r}\,')\}}{|\vec{r} - \vec{r}\,'|} d\vec{r}\, d\vec{r}\,' = A_{rqts} \tag{11}$$

E^{S} can be written in the following form, in accord with (3), (7), and (10):

$$2E^{\mathrm{S}} = \sum_{qrst} a^\dagger_q a_r a^\dagger_s a_t A_{qrst} - 2\sum_{qr} a^\dagger_q q_r \sum_{\bar{s}} A_{qr\bar{s}\bar{s}} + \sum_{\bar{s}\bar{t}} A_{\bar{s}\bar{s}\bar{t}\bar{t}}.$$

To a first approximation, $2E^{\mathrm{S}}$ is the diagonal element of the operator E^{S}, which corresponds to the occupation state $\ldots N_q \ldots$ under consideration. Since this is already proportional to e^2, we can limit ourselves to this approximation. Among the operators $a^\dagger_q a_r a^\dagger_s a_t$ and $a^\dagger_q a_r$ only the following combinations have diagonal elements:

$$a^\dagger_q a_q a^\dagger_r a_r = N_q N_r,$$
$$a^\dagger_q a_r a^\dagger_r a_q = N_q(1 - N_r),$$
$$a^\dagger_q a_q = N_q,$$

which follows easily from (9).

We thus obtain for the diagonal element of E^{S}:

$$2E^{\mathrm{S}} = \sum_{qr} N_q N_r A_{qqrr} - 2\sum_{r\bar{s}} N_r A_{rr\bar{s}\bar{s}} + \sum_{\bar{r}\bar{s}} A_{\bar{r}\bar{r}\bar{s}\bar{s}} + \sum_{qr} N_q(1 - N_r)A_{qrrq}. \tag{12}$$

We calculate the Coulomb energy of the 'vacuum' (that is $N_{\overset{+}{q}} = 0$, $N_{\bar{q}} = 1$ for all q). On this basis, we obtain:

$$2E^{\mathrm{S}}_{\mathrm{vac}} = \sum_{\overset{+}{q}\bar{r}} A_{\overset{+}{q}\bar{r}\bar{r}\overset{+}{q}}.$$

The total energy of the electron in state $\overset{+}{q}_0$ with occupied negative energy levels becomes

$$2E^{\mathrm{S}}_{\mathrm{vac}+1} = A_{\overset{+}{q}_0\overset{+}{q}_0\overset{+}{q}_0\overset{+}{q}_0} + \sum_{\overset{+}{r}\neq\overset{+}{q}_0} A_{\overset{+}{q}_0\overset{+}{r}\overset{+}{r}\overset{+}{q}_0} + \sum_{\bar{r}\overset{+}{s}} A_{\bar{r}\overset{+}{s}\overset{+}{s}\bar{r}} - \sum_{\bar{r}} A_{\bar{r}\overset{+}{q}_0\overset{+}{q}_0\bar{r}},$$

so that, for the electrostatic self-energy of an electron in the state q_0, we obtain the difference of the two values:

$$2E^{\mathrm{S}} = \sum_{\overset{+}{r}} A_{\overset{+}{q}_0\overset{+}{r}\overset{+}{r}\overset{+}{q}_0} - \sum_{\bar{r}} A_{\overset{+}{q}_0\bar{r}\bar{r}\overset{+}{q}_0}. \tag{13}$$

The electrostatic self-energy according to the usual Dirac theory without occupation of negative energy states is obtained from (12) by setting $\tilde{\rho} = 0$ and $N_q = 0$ for all $\overset{+}{q}$ and $\bar{q}$ except for $\overset{+}{q}_0 = 1$. One then obtains

$$2E^{\mathrm{S}} = \sum_{\overset{+}{r}} A_{\overset{+}{q}_0\overset{+}{r}\overset{+}{r}\overset{+}{q}_0} + \sum_{\bar{r}} A_{\overset{+}{q}_0\bar{r}\bar{r}\overset{+}{q}_0}. \tag{14}$$

The sums can be calculated in the following fashion: With the more complete notations p^k of the states, we obtain for $A_{q_0 r r q_0}$ according to (11) and (5)

$$A_{p_0^k p^l p^l p_0^k} = \frac{e^2}{h^6}\{u^{k*}(p_0)u^l(p)\}\{u^{l*}(p)u^k(p_0)\}$$

$$\cdot \int \exp\frac{\left[\frac{\mathrm{i}}{\hbar}(\vec{p} - \vec{p}_0, \vec{r} - \vec{r}')\right]}{|\vec{r} - \vec{r}'|}\mathrm{d}\vec{r}\,\mathrm{d}\vec{r}.$$

The integral has the value

$$\frac{h^2}{\pi}\frac{1}{(\vec{p} - \vec{p}_0)^2}\cdot V,$$

where V is the total volume $\int \mathrm{d}\vec{r}$ under consideration.

We now take into account the normalization of the eigenfunctions and – as mentioned under (6) – reduce the result to a single electron through multiplication by h^3/V, and we obtain:

$$A_{p_0^k p^l p^l p_0^k} = \frac{e^2}{\pi h}\frac{\{u^{k*}(p_0)u^l(p)\}\{u^{l*}(p)u^k(p_0)\}}{(\vec{p} - \vec{p}_0)^2},$$

and thus for (13)

$$\frac{2\pi h}{e^2}E^{\mathrm{S}} = \left(\sum_{l=1,2} - \sum_{l=3,4}\right)\int \frac{\{u^{k*}(p_0)u^l(p)\}\{u^{l*}(p)u^k(p_0)\}}{(\vec{p} - \vec{p}_0)^2}\mathrm{d}\vec{p}.$$

Evaluation of the sum over l is greatly facilitated by simultaneously summing also over the values $k = 1$ and 2, which yields only a factor of 2, since the energy of a free electron cannot depend on the spin direction. It is always very convenient, when summing over the two spin states, to use the following relation:

$$\frac{(\vec{\alpha},\vec{p})+\beta mc}{P}\,u^k(p)=\begin{cases}+\,u^k(p) & \text{for } k=1,2\\ -\,u^k(p) & \text{for } k=3,4,\end{cases}$$

where below the following shall also always hold:

$$P=+(p^2+m^2c^2)^{1/2}. \tag{15}$$

The E^{S} can be written in a form where the summation extends over all four indices:

$$\frac{2\pi h}{e^2}\,E^{\mathrm{S}}$$

$$=\sum_{k,l}^{1\ldots 4}\frac{1}{4}\int\frac{\left\{u^{k*}(p_0)\dfrac{(\vec{\alpha},\vec{p})+\beta mc}{P}\,u^l(p)\right\}\left\{u^{*l}(p)\left[1+\dfrac{(\vec{\alpha},\vec{p}_0)+\beta mc}{P_0}\right]u^k(p)\right\}}{(\vec{p}-\vec{p}_0)^2}\,\mathrm{d}\vec{p}.$$

One can use the convenient relation

$$\sum_{k,l}^{1..4}\{u^{k*}(p_0),Ou^l(p)\}\{u^{l*}(p),O'u^k(p)\}=Sp(O\cdot O')$$

where O and O' are thought of as four-row matrices in spin space.[11] This relation follows from the orthogonality relations (6). One thus becomes independent of the special representation of the $u^k(p)$.

From the relation

$$Sp[(\vec{a},\vec{p})+\beta mc][P_0+(\vec{\alpha},\vec{p}_0)+\beta mc]=4(\vec{p},\vec{p}_0)+4m^2c^2$$

there immediately follows

$$\frac{2\pi h}{e^2}\,E^{\mathrm{S}}=\int\frac{(\vec{p},\vec{p}_0)+m^2c^2}{PP_0(\vec{p}-\vec{p})^2}\,\mathrm{d}\vec{p}.$$

After integration over the directions of $\vec{p}$, this yields an integral over the magnitudes of $|\vec{p}|=p$:

$$E^{\mathrm{S}}=\frac{e^2}{hP_0}\int_0^\infty\mathrm{d}p\left[\frac{p}{2p_0P}\,(m^2c^2+\tfrac{1}{2}(p^2+p_0^2))\log\frac{(p+p_0)^2}{(p-p_0)^2}-\frac{p^2}{P}\right].$$

If we expand the integrand for $p>p_0$ in powers of p, we obtain:

$$E^{\mathrm{S}}=\frac{e^2}{hP_0}\,(2m^2c^2+p_0^2)\cdot\int\frac{\mathrm{d}p}{p}+\cdots,$$

where we omit terms of higher order in $1/p$, which yield finite contributions upon integration. The electrostatic self-energy thus diverges *logarithmically* in the 'hole' theory.

By comparison, let us specify the self-energy of the electron according to the Dirac single-electron theory, as calculated according to (14):

$$E^{\mathrm{S}}=\frac{e^2}{2\pi h}\int\frac{\mathrm{d}\vec{p}}{(\vec{p}-\vec{p})^2}=\frac{e^2}{h}\int_0^\infty\frac{\mathrm{d}p\cdot p}{2p_0}\log\frac{(p+p_0)^2}{(p-p_0)^2}=\frac{e^2}{h}\int\mathrm{d}p+\cdots,$$

which diverges linearly.

[11] $Sp(O,O')$ signifies the trace of the matrix product $O\cdot O'$.

(b) *The electrodynamic self-energy*

We now arrive at the calculation of E^{D}. Here we expand the vector potential $\vec{A}_{\mathrm{tr}}$ in powers of e:

$$\vec{A}_{\mathrm{tr}} = \vec{A}^0_{\mathrm{tr}} + \vec{A}^1_{\mathrm{tr}} + \cdots.$$

The portion $\vec{A}^0_{\mathrm{tr}}$ which is independent of e may not be set equal to zero, although it naturally vanishes according to classical electrodynamics. In fact, however, it represents the zero-point fluctuations of the field strengths which exist in empty space. In general, one can take into account the effect of this zero-point field strength, which causes spontaneous emission, by writing down the light field existing in space in the following fashion in a Fourier decomposition:

$$\vec{A}^0_{\mathrm{tr}} = \sum_j^{1,2} \int \mathrm{d}\vec{k} \left(\vec{A}^+_j(\vec{k}) \exp\left[\frac{\mathrm{i}}{\hbar}(\vec{k}, \vec{r}) + \frac{\mathrm{i}c}{\hbar} kt\right] + \vec{A}^-_j(\vec{k}) \exp\left[-\frac{\mathrm{i}}{\hbar}(\vec{k}, \vec{r}) - \frac{\mathrm{i}c}{\hbar} kt\right]\right). \tag{16}$$

The index j designates the two polarization directions ($k = |k|$). As a consequence of the time periodicity of $\vec{A}_{\mathrm{tr}}$, quadratic expressions of the form $\vec{A}^+_j(k)\vec{A}^-_j(k)$ or $\vec{A}^-_j(k)\vec{A}^+_j(k)$ will always occur in $\vec{A}_{\mathrm{tr}}$ in the results. According to classical electrodynamics, we have:

$$\vec{A}^+_j(\vec{k})\vec{A}^-_j(\vec{k}) + \vec{A}^-_j(\vec{k})\vec{A}^+_j(\vec{k}) = \frac{h^2}{\pi k^2} S_j(\vec{k}), \tag{17}$$

where $S_j(\vec{k})\,\mathrm{d}\vec{k}$ is the energy density in the radiation with the momentum vector $\vec{k}$ and the polarization j.

To satisfy quantum theory, one must at this point take into account the non-commutability of the $\vec{A}_j(k)$. If, in the expression for the charge density, one specifies the sequence $e\psi^+\psi$ (instead of $e\psi\psi^+$), one should set

$$\left.\begin{aligned} 2\vec{A}^+_j(\vec{k})\vec{A}^-_j(\vec{k}) &= \frac{h^2}{\pi k^2}\left(S_j(\vec{k}) + \frac{ck}{h^3}\right), \\ 2A^-_j(\vec{k})A^+_j(\vec{k}) &= \frac{h^2}{\pi k^2} S_j(\vec{k}). \end{aligned}\right\} \tag{18}$$

Whereas the classical expression (17) yields only those transitions of an atomic system that are forced by the external radiation field, the addition ck/h^3 yields the additional spontaneous emission transitions, since the combination $\vec{A}^+_j\vec{A}^-_j$ occurs only for emission, $\vec{A}^-_j\vec{A}^+_j$ only for absorption.[12]

$\vec{A}^1_{\mathrm{tr}}$ is calculated according to the formula

$$\vec{A}^1_{\mathrm{tr}}(\vec{r}) = \int \frac{[\vec{i}^0_{\mathrm{tr}}(\vec{r}')]}{|\vec{r} - \vec{r}'|} \mathrm{d}\vec{r}', \tag{19}$$

where the square brackets indicate that the current $\vec{i}^0_{\mathrm{tr}}$ is to be taken retarded. The designation $\vec{i}^0_{\mathrm{tr}}$ indicates that, in this approximation, only that current is to be considered which is unperturbed by the potentials. In its time dependence, according to (7) and (8), this is:

[12] This prescription is treated in more detail, for example, in the article by W. Pauli in Geiger-Scheels *Handb. d. Phys.* page 210.

$$\vec{i}^{\,0}_{\rm tr}(\vec{r}\,') = \frac{e}{h^3}\sum_{k,l}\int\int {\rm d}\vec{p}\,{\rm d}\vec{p}\,' a^{k\dagger}(p')a^l(p)\{u^{k*}(p'), \vec{\alpha}_s u^l(p)\}$$
$$\cdot \exp\left[\frac{\rm i}{\hbar}(\vec{p}-\vec{p}\,', \vec{r}\,')\right]\cdot\exp\left[\frac{\rm i}{\hbar}(E^l(p)-E^k(p'))t\right]. \qquad (20)$$

Here, $\vec{\alpha}_s$ is a component of $\vec{\alpha}$ perpendicular to $\vec{p}-\vec{p}\,'$, which must be taken to assure that $\vec{i}^{\,0}_{\rm tr}$ is divergence-free. By performing the integration over $\vec{r}\,'$ and after introducing the new variables $\vec{k}=\vec{p}\,'-\vec{p}$, there follows from this:

$$A^1_{\rm tr}(\vec{r}\,')$$
$$= \frac{ec}{2\pi h}\sum_{k,l}\int {\rm d}\vec{p}\int {\rm d}\vec{k}\, a^{k\dagger}(\vec{p}+\vec{k})a^l(p)\{u^{k*}(\vec{p}+\vec{k}), \vec{\alpha}_s u^l(\vec{p})\}$$
$$\cdot\frac{1}{k}\left(\frac{1}{E^l(p)-E^k(\vec{p}+\vec{k})+ck} - \frac{1}{E^l(p)-E^k(\vec{p}+\vec{k})-ck}\right) \qquad (21)$$
$$\cdot\exp\left[\frac{\rm i}{\hbar}(\vec{k},\vec{r}\,')\right]\cdot\exp\left[\frac{\rm i}{\hbar}(E^l(p)-E^k(\vec{p}+\vec{k}))t\right]$$

Just like the potential $\vec{A}_{\rm tr}$, we expand the current density $\vec{i}_{\rm tr}$:

$$\vec{i}_{\rm tr} = \vec{i}^{\,0}_{\rm tr} + \vec{i}^{\,1}_{\rm tr} + \cdots,$$

where $\vec{i}^{\,0}_{\rm tr}$ is proportional to e and is given by (20); $\vec{i}^{\,1}_{\rm tr}$ is the additional current density caused by $\vec{A}^0_{\rm tr}$, which we can calculate by a perturbation procedure. One thus obtains

$$\vec{i}^{\,1}_{\rm tr} =$$
$$-\frac{e^2}{h^3}\sum_j\sum_{kl}\int {\rm d}\vec{p}\int {\rm d}\vec{k}\, a^{k\dagger}(\vec{p})a^l(\vec{p}+\vec{k})a^{l\dagger}(\vec{p}+\vec{k})a^k(\vec{p})\{u^{k*}(\vec{p}), \vec{\alpha}_s u^l(\vec{p}+\vec{k})\}$$
$$\cdot\left[\frac{\{u^{l*}(\vec{p}+\vec{k}), (\vec{\alpha}, \vec{A}^+_j(\vec{k}))u^k(\vec{p})\}}{E^k(p)-E^l(\vec{p}+\vec{k})+ck}\exp\left(\frac{\rm i}{\hbar}(\vec{k},\vec{r}\,')\right)\right.$$
$$\left.+\frac{\{u^{l*}(\vec{p}+\vec{k}), (\vec{\alpha}, A^+_j(\vec{k}))u^k(\vec{p})\}}{E^k(p)-E^l(\vec{p}+\vec{k})+ck}\right]\exp\left(\frac{\rm ic}{\hbar}kt\right)$$
$$+ \text{corresponding terms with } A^-_j(\vec{k}) + \text{time-independent terms.} \qquad (22)$$

The added terms that have been indicated play no role in the final result.

For the total electrodynamic energy we thus obtain:

$$E^{\rm D} = -\frac{1}{2}\int(\vec{i}^{\,0}_{\rm tr} - \vec{i}^{\,1}_{\rm tr} - \vec{\vec{i}}, \vec{A}^0_{\rm tr} + \vec{A}^1_{\rm tr} - \vec{\vec{A}}_{\rm tr}).$$

One can easily convince oneself that the potential $\vec{\vec{A}}_{\rm tr}$, originating from the fully occupied negative states, is equal to zero. Furthermore, due to the time periodicity, the combinations $(\vec{i}^{\,0}_{\rm tr} + \vec{\vec{i}}_{\rm tr}\vec{A}^0_{\rm tr})$ and $(\vec{i}A^1_{\rm tr})$ vanish. After omitting the term $(\vec{i}^{\,1}_{\rm tr}, \vec{A}^1_{\rm tr})$, which is proportional to e^3, there remains:

$$E^{\rm D} = -\frac{1}{2}\int(\vec{i}^{\,0}_{\rm tr}, \vec{A}^1_{\rm tr})\,{\rm d}\vec{r} - \frac{1}{2}\int(\vec{i}^{\,1}_{\rm tr}, \vec{A}^0_{\rm tr})\,{\rm d}\vec{r}.$$

The four-vectors occuring here are given in (16), (20), (21) and (22). When calculating the $\int(\vec{i}^{\,1}_{\rm tr}\vec{A}^0_{\rm tr})\,{\rm d}\vec{r}$ one must insert the following according to (18):[13]

[13] At this point, it appears as if the result were to depend on the sequence $\int(\vec{i}^{\,1}_{\rm tr}\vec{A}^0)\,{\rm d}\vec{r}$ or $\int(\vec{A}^0\vec{i}^{\,1}_{\rm tr})\,{\rm d}\vec{r}$. However, it is easily seen that such a commutation also changes the sequence of the operators $a(p)$ in $\int(\vec{i}^{\,0}_{\rm tr}\vec{A}^1)\,{\rm d}\vec{r}$ in such a fashion that the same result finally appears.

$$A_j^+(\vec{k})A_j^-(\vec{k}) = A_j^+(-\vec{k})A_j^-(-\vec{k}) = \frac{c}{2\pi kh}.$$

One then obtains

$$E^{\mathrm{D}} = \frac{e^2}{2\pi h}\sum_{kl}\int \mathrm{d}\vec{p}\int\frac{\mathrm{d}\vec{k}}{k}N^k(\vec{p})(1 - N^l(\vec{p}+\vec{k}))$$
$$\cdot\frac{\{u^{k*}(\vec{p}), \vec{\alpha}_s u^l(\vec{p}+\vec{k})\}\{u^{*l}(\vec{p}+\vec{k}), \vec{\alpha}_s u^k(\vec{p})\}}{E^k(p) - E^l(\vec{p}+\vec{k}) - ck}. \tag{23}$$

Here, by means of (9), the occupation numbers $N^k(\vec{p})$ have already been inserted in place of the operators $a^k(p)$. $\vec{\alpha}_S$ signifies the component of $\vec{\alpha}$ perpendicular to $\vec{k}$. The multiplication by h^3/V also has already been done (see note after formula (6′)).

After reduction to the case of one electron, (23) is identical to the expression derived by Waller on the basis of the Dirac radiation theory. It is easily seen that Waller's derivation of the multi-electron problem also yields the factor $N^k(\vec{p})(1 - N^1(\vec{p}+\vec{k}))$.

If we specialize the expression (23) to the case calculated by Waller, namely the case of electron with momentum p_0 ($N^k(\vec{p}) = \delta(p - p_0)$, $k = 1$ or 2), we obtain:

$$E^{\mathrm{D}} = J_+^k(p_0) + J_-^k(p_0)$$
$$J_\pm^k(\vec{p}_0) = \frac{e^2c}{2\pi h}\sum_t\int\frac{\mathrm{d}\vec{k}}{k}\frac{\{u^{k*}(\vec{p}_0)\vec{\alpha}_s u^l(\vec{p}+\vec{k})\}\{u^{*l}(\vec{p}+\vec{k})\vec{\alpha}_s u^k(\vec{p})\}}{E^k(\vec{p}) - E^l(\vec{p}+\vec{k}) - ck},$$

where, in J_+^k, one must sum over the values $l = 1, 2$, and in J_-^k one must sum over the values $l = 3, 4$.

For the 'vacuum', ($N^k(p) = 1$ for $k = 3, 4$; $N^k(\mathrm{p}) = 0$ for $k = 1, 2$), one obtains

$$E_{\mathrm{vac}}^{\mathrm{D}} = \sum_{k=3,4}\int J_-^k(\vec{p})\,\mathrm{d}\vec{p}.$$

For an electron in state $\vec{p}_0^k$, $k = 1$ or 2, with occupied negative states ($N^k(\vec{p}) = 1$ for $k = 3, 4$; $N^k(\vec{p}) = \delta(\vec{p} - \vec{p}_0)$ for $k = 1$ or 2) one obtains:

$$E_{\mathrm{vac}+1}^{\mathrm{D}} = J_+^k(\vec{p}_0) + E_{\mathrm{vac}}^{\mathrm{D}}$$
$$-\frac{e^2c}{2\pi h}\sum_{l=3,4}\int\frac{\mathrm{d}\vec{k}}{k}\frac{\{u^{l*}(\vec{p}_0 - \vec{k}), \vec{\alpha}_s u^k(\vec{p}_0)\}\{u^{k*}(\vec{p}_0), \vec{\alpha}_s u^l(\vec{p}_0 - \vec{k})\}}{E^l(\vec{p} - \vec{k}) - E^k(p) - ck}.$$

By introducing a new integration variable, it is easily seen that the third term is equal to J_-^k. We thus obtain for the self-energy of an electron in state p_0^k, in the 'hole theory', after subtracting the 'vacuum' energy:

$$E^{\mathrm{D}} = J_+^k(p_0) - J_-^k(p_0)$$

The calculation of the $J^k(p)$ is done with the same calculational methods as the calculation of the electrostatic energy. One thus obtains:

$$J_+^k(\vec{p}) = \frac{e^2}{2\pi h}\int\frac{\mathrm{d}\vec{k}}{k}\frac{PP_+ - \frac{1}{k^2}(\vec{k},\vec{p})^2 - (\vec{k},\vec{p}) - m^2c^2}{PP_+(P - P_+ - k)},$$

$$J_-^k(\vec{p}) = \frac{e^2}{2\pi h}\int\frac{\mathrm{d}\vec{k}}{k}\frac{PP_+ + \frac{1}{k^2}(\vec{k},\vec{p})^2 + (\vec{k},\vec{p}) + m^2c^2}{PP_+(P + P_+ - k)}, \text{ for } k = 1, 2,$$

where

$$P = +(p^2 + m^2c^2)^{1/2},\ P_+ = +[(\vec{p} + \vec{k})^2 + m^2c^2]^{1/2}.$$

Integration over the directions of $\vec{k}$ yields:

$$J^k_\pm(p) = \frac{e^2}{hP}\Bigg[\int_0^\infty \mathrm{d}k\,\frac{m^2c^2}{p}\log\frac{P-k\mp P_+}{P-k\mp P_-} - \int_0^\infty \mathrm{d}k(P-k)$$
$$\mp\frac{1}{4}\int_0^\infty \frac{\mathrm{d}k}{k^2 p}(P^3 + P^2k + Pk^2 + k^3)(P_+ - P_-)$$
$$\mp\frac{1}{12}\int_0^\infty \frac{\mathrm{d}k}{k^2 p}(P-k)(P_+^3 - P_-^3)\Bigg],$$

where the symbol $P_\pm$ has the following meaning:

$$P_\pm = [(\vec{p} \pm \vec{k})^2 + m^2c^2]^{1/2}.$$

From this, one sees that $J^k_+ + J^k_-$ yields a simple expression, namely

$$J^k_+ + J^k_- = \frac{e^2}{h}\left[\frac{m^2c^2}{pP}\log\frac{P+p}{P-p} - 2\right]\int_0^\infty \mathrm{d}k + \frac{2e^2}{hP}\int_0^\infty k\,\mathrm{d}k,$$

which represents the self-energy of a single free electron with momentum p and which is identical to the expression previously derived by Waller. The difference $J^k_+ - J^k_-$, which represents the self-energy of an electron with occupied negative energy states, is not so clear. We therefore give only the terms which, in an expansion of the integrand in powers of k, lead to divergent integrals for $k < p$:

$$E^{\mathrm{D}} = -\frac{11}{6}\frac{e^2}{hP}p^2\int_0^\infty \frac{\mathrm{d}k}{k} - \frac{e^2}{h}\frac{m^2c^2}{pP}\log\frac{P+p}{P-p}\int_0^\infty \mathrm{d}k - \frac{2e^2}{hP}\int_0^\infty k\,\mathrm{d}k$$
$$+ \text{finite terms}.$$

From this one can also see that the degree of divergence of the self-energy does not become smaller through the occupation of negative states.[14]

Finally, I wish to express my sincere gratitude to Professor Pauli for stimulating this work as well as for constant help in its performance. I likewise owe special thanks to Professor Niels Bohr for the discussion of the theoretical foundations.

Zürich, Physical Institute of the Federal Technical College.

Correction to the paper: The self-energy of the electron

Zeitschrift für Physik, 90: 817–18 (1934). Received 20 July 1934.

On [p. 166] of the paper cited above, there is a computational error which has seriously garbled the results of the calculation for the electrodynamic self-energy of the electron

[14] See note 4 [p. 157].

according to the Dirac hole theory. I am greatly indebted to Mr Furry (University of California, Berkeley) for kindly pointing this out to me.

The degree of divergence of the self-energy in the hole theory is *not*, as asserted in [the preceding paper], just as great as in the Dirac one-electron theory, but the divergence is only logarithmic. The expression for the electrostatic and electrodynamic parts of the self-energy E of an electron with momentum p now correctly reads, in the notations used in [the preceding paper]:

$$E = E^{\mathrm{S}} + E^{\mathrm{D}},$$

$$E^{\mathrm{S}} = \frac{e^2}{h(m^2c^2 + p^2)^{1/2}} (2m^2c^2 + p^2) \int_{k_0}^{\infty} \frac{\mathrm{d}k}{k} + \text{finite terms},$$

$$E^{\mathrm{D}} = \frac{e^2}{h(m^2c^2 + p^2)^{1/2}} (m^2c^2 - \tfrac{4}{3} p^2) \int_{k_0}^{\infty} \frac{\mathrm{d}k}{k} + \text{finite terms}.$$

For comparison, we cite the expressions obtained on the basis of the single-electron theory:

$$E^{\mathrm{S}} = \frac{c^2}{h} \int_0^{\infty} \mathrm{d}k + \text{finite terms},$$

$$E^{\mathrm{D}} = \frac{e^2}{h} \left[\frac{m^2c^2}{p(m^2c^2 + p^2)^{1/2}} \log \frac{(m^2c^2 + p^2)^{1/2} + p}{(m^2c^2 + p^2)^{1/2} - p} - 2 \right] \int_0^{\infty} \mathrm{d}k + \frac{2e^2}{h(m^2c^2 + p^2)^{1/2}} \int_0^{\infty} k \, \mathrm{d}k.$$

The computational error arose in the transformation of the electrodynamic portion E^{D} for the case of the hole theory:

$$E^{\mathrm{D}} = J_+^k(\vec{p}) - J_-^{\prime k}(\vec{p}), \quad k = 1 \text{ or } 2,$$

where $J_+^k(\vec{p})$ is defined on [p. 166] whereas

$$J_-^{\prime k}(\vec{p}) = -\frac{e^2}{2\pi h} \int \frac{\mathrm{d}\vec{k}}{k} \frac{PP_+ + \frac{1}{k^2}(\vec{k}\vec{p})^2 + (\vec{k}\vec{p}) + m^2c^2}{PP_+(P + P_+ + k)}$$

and is not equal to the quantity $J_-^k(\vec{p})$, from which it differs only by a sign. Likewise, one must set

$$E_{\mathrm{vac}}^{\mathrm{D}} = \sum_{k=1,2} \int J_-^{\prime k}(\vec{p}) \, \mathrm{d}\vec{p}$$

for the self-energy of the vacuum.

As a consequence of the new result, the question raised in note 4 of the paper requires a new examination, whether the Wentzel method,[15] to avoid the infinite self-energy by suitable limiting processes, might not still lead to the objective in the hole theory.

[15] G. Wentzel, *ZS. f. Phys*, **86**, 479, 635, 1933.

6 Remarks on the Dirac theory of the positron

W. HEISENBERG

Zeitschrift für Physik, 90: 209–31 (1934). Received 21 June 1934.

I. Intuitive theory of matter waves: 1. The inhomogeneous differential equation of the density matrix, 2. The conservation laws, 3. Applications (polarization of the vacuum). II. Quantum theory of wave fields: 1. Setting up the basic equations, 2. Applications (the self-energy of light quanta).

The intention of the present paper[1] is to build the Dirac theory of the positron[2] into the formalism of quantum electrodynamics. A requirement here is that the symmetry of nature in the positive and negative charges should from the very beginning be expressed in the basic equations of theory, and further that, except for the divergences which are caused by the well-known difficulties of quantum electrodynamics, no new infinities appear in the formalism, i.e. that the theory provides an approximation method for treating the group of problems which could also be treated according to the previous quantum electrodynamics. The latter postulate distinguishes the present attempt from the investigations of Fock[3], Oppenheimer and Furry[4], and Peierls[5], which it otherwise resembles; it is here rather closely connected with a paper by Dirac[6]. Compared to Dirac's treatment, this paper emphasizes the significance of the conservation laws for the total system of radiation and matter, and the necessity of formulating the basic equations of the theory in a manner extending beyond the Hartree approximation.

I Intuitive theory of matter waves

1 *The inhomogeneous differential equation of the density matrix*

The most important results of the above-mentioned Dirac paper will first be briefly reviewed: A quantum mechanical system of many electrons, which fulfils the Pauli principle, and which moves without mutual interaction in a prescribed force field, can be characterized by a 'density matrix':

$$(x't'k'|R|x''t''k'') = \sum_n \psi_n^*(x't'k')\psi_n(x''t''k''), \tag{1}$$

[1] This paper arose from discussions which I had with Pauli, Dirac, and Weisskopf, partly in writing and partly verbally, and for which I thank them sincerely.

[2] e.g.: P. A. M. Dirac, *The Principles of Quantum Mechanics*, p. 255, Oxford (1930).

[3] V. Fock, *C.R. Leningrad* (N.S.) 1933, pp. 267–71, no. 6.

[4] W. H. Furry and I. R. Oppenheimer, *Phys. Rev.* **45**, 245, 1935.

[5] R. Peierls, in press. [*Proc. Roy. Soc. (A)* **146**: 420–41 (1934).]

[6] P. A. M. Dirac, *Proc. Cambr. Phil. Soc.* **30**, 150, 1934 (below always cited as *loc. cit.*).

where $\psi_n(x't'k')$ signify the normalized eigenfunctions of states that are occupied with an electron, and $x't'k'$, respectively $x''t''k''$, are position, time, and spin variables. [Heisenberg and Hans Euler noted several errors in the present paper in their sequel paper 'Conclusions from Dirac's theory of the positron', *Zeitschrift für Physik*, 98: 714–32 (1936). They wrote that in Eq. (1) 'the single and double primes are erroneously interchanged on the right hand side'.] The density matrix yields all the physically important properties of the quantum mechanical system, such as charge density, current density, energy density, etc. However, this also only holds in the approximation in which one neglects the interaction of the electrons, i.e. in which the typically quantum theoretical non-intuitive traits of the event do not occur; the density matrix thus imparts an intuitive correspondence-like picture of the actual process – similar to that of the classical-mechanical atomic models; the requirement that the ψ_n in (1) should be normalized, which according to Dirac is also expressed in the form (for $t' = t''$)

$$R^2 = R, \tag{2}$$

can be set in parallel with the quantum conditions of the previous semi-classical theory.

The time change of the density matrix is determined by the Dirac differential equation:

$$\mathcal{H}R = \left[i\hbar\frac{\partial}{c\partial t'} + \frac{e}{c}A_0(x') + \alpha_s\left(i\hbar\frac{\partial}{\partial x'_s} - \frac{e}{c}A_s(x')\right) + \beta mc\right]R = 0. \tag{3}$$

From now on, the following notation will always be used:

$$\left.\begin{aligned}
&\textit{Coordinates:}\\
&ct' = x'_0 = -x^{0'},\ x'_i = x^{i'},\ x'_\lambda - x''_\lambda = x_\lambda,\ \frac{x'_\lambda + x''_\lambda}{2} = \xi_\lambda.\\
&\textit{Potentials:}\\
&A_0 = -A^0,\ A_i = A^i,\\
&\textit{Field strengths:}\\
&\frac{\partial A^\mu}{\partial \xi_\nu} - \frac{\partial A^\nu}{\partial \xi_\mu} = F^{\nu\mu},\ F^{0s} = -F_{0s}.\\
&(F^{01}, F^{02}, F^{03}) = \boldsymbol{E},\ (F^{23}, F^{31}, F^{12}) = \boldsymbol{H}.\\
&\textit{Spin matrices:}\\
&\alpha^0 = 1,\ \alpha_0 = -1,\ \alpha^i = \alpha_i.
\end{aligned}\right\} \tag{4}$$

Greek indices always run from 0 to 3, Latin ones from 1 to 3. The indices will be raised and lowered according to the usual formulas of relativity theory. One always sums over indices which appear doubly. Since the α^ν do not transform simply as vectors, the chosen notation is valuable only as a suitable abbreviation for these quantities. Equation (3), for example, now assumes the form:

$$\left\{\alpha^\lambda\left[i\hbar\frac{\partial}{\partial x'_\lambda} - \frac{e}{c}A^\lambda(x')\right] + \beta mc\right\}R = 0.$$

If, as the Dirac hole theory requires, all states of negative energy are occupied except for a finite number, and also if only a finite number of states of positive energy are occupied, the matrix R becomes singular on the light cone defined by

$$x_\rho x^\rho = 0. \tag{5}$$

According to Dirac, it is then appropriate to consider, in place of the matrix R, the new matrix[7]

$$R_S = R - \tfrac{1}{2}R_F, \tag{6}$$

where R_F designates the value of R for the state of the system at which each electron level is occupied. For $t' = t''$, R_F goes over into the Dirac δ-function of the variables $x'k'$, $x''k''$, as is easily demonstrated. The matrix R_S already has symmetry with respect to the sign of the charge, which will be important later on in the formalism: By adding $\frac{1}{2}R_F$, it goes over into the matrix R, which corresponds to the 'hole' theory; by subtracting $\frac{1}{2}R_F$, it goes over into the negative density matrix of a distribution where the states of positive energy are occupied and those of negative energy are empty; exchange of the points $x't'k'$ and $x''t''k''$ in R_S, and a sign change of R_S, are equivalent to a sign change of the electron charge. The singularity of the matrix R_S along the light cone has been investigated by Dirac; one can represent the matrix in the form

$$(x'k'|R_S|x''k'') = u\frac{\alpha^\rho x_\rho}{(x^\lambda x_\lambda)^2} - \frac{v}{x^\lambda x_\lambda} + w \log|x^\lambda x_\lambda| \tag{7}$$

where

$$u = -\frac{\mathrm{i}}{2\pi^2}\exp\left(\frac{ei}{c\hbar}\int_{P'}^{P''} A^\lambda \,\mathrm{d}x_\lambda\right). \tag{8}$$

(The integral is to be taken along the straight line from P' to P''.) [Eq. (8) 'erroneously contains the negative sign in the exponent', Heisenberg and Euler, *loc. cit.*]

The quantity w is uniquely specified by a differential equation; v is determined only up to an additive term of the form $x^\lambda x_\lambda \cdot g$. Generally one proceeds from the density matrix R to the charge density, current density, etc., by making the following formulation e.g. for the charge density:

$$\rho(x) = e\sum_k (xk|R|xk) \tag{9}$$

and correspondingly for the other physical quantities. Now this conclusion is evidently incorrect because of the singularity of the matrix R. For example, if no external field is present, only the deviation of the density matrix from the matrix of the state where all levels of negative energy are filled up will contribute to the charge density and the current density. Consequently, according to Dirac, one will have to subtract from the density matrix another density matrix which is determined uniquely by the external fields, in order to obtain the 'real' density matrix – we shall call it $(x'k'|r|x''k'')$. This will determine the charge density, current density, energy density, etc. in correspondence with Equation (9). We set

$$r = R_S - S \tag{10}$$

where S is supposed to be a function of $x'_\lambda k'$ and $x''_\lambda k''$, which is uniquely determined by the potentials A^λ.

In place of the differential equation (3), there now appears the equation

$$\mathcal{H}r = -\mathcal{H}S. \tag{11}$$

The right side is a function of the electromagnetic field, a function which must still be determined more precisely; the original homogeneous Dirac Equation (3) is accordingly replaced by the inhomogeneous Equation (11). Such an equation is the natural

[7] Twice the matrix R_S is the matrix which Dirac denotes by R_1.

expression of the fact that matter can be created and destroyed; the type of generation and annihilation is specified by the mathematical form of the quantity $\mathscr{H}S$. If no external fields are present, S is supposed to be given by the value of R_S for the distribution where all states of negative energy are occupied; for we assume that, in the field-free vacuum, the matrix r vanishes everywhere. The quantity of matter which is created as a whole when an external field is switched on and switched off again can be determined without a more detailed determination of S if external fields are present. For if R_S (and thus r) was known before switching on any fields, the value of R_S after the field has been switched off again can be determined from Equation (3). After the field has been switched off, S has its original value again, however, and thus r can also be calculated. But, inversely, the results concerning the creation of matter when fields are switched on and off can give general reference points concerning the form of the right side of (11) in the presence of fields. For example, a simple perturbation calculation shows that the total amount of matter created during the switching on and switching off is generally already infinite when the time differential quotient of the electric or magnetic field strength was discontinuous at some time during the switch-on and switch-off process, and especially so when the field strength or potentials themselves were discontinuous; from this one can conclude that the right side of (11) must contain not only the potentials and field strength but also their first and second derivatives.

Dirac (*loc. cit.*) undertakes the determination of S in the presence of external fields in such a way as to describe a particular mathematical method which yields one-by-one the singular portions of the matrix R_S; Dirac identifies the sum of the singular portions obtained in this way with S. The mathematical method chosen by Dirac, however, does not yield the value of S defined above in the force-free case, but rather a value which differs from it by a matrix that is regular on the light cone. Accordingly, it is not possible to specify uniquely the inhomogeneity in (11) from formal arguments below; nevertheless, by taking into account the conservation laws of charge, energy, and momentum, the possibilities for S can be restricted insofar that a particular value can be distinguished as the simplest assumption. That value of S which holds in the absence of external forces and potentials (compare above) and which is calculated in Dirac, *loc. cit.*, Equations (20) through (22), we designate as S_0. If there are no fields present, but potentials do occur in Equation (3), whose curl vanishes, S_0 is to be replaced by

$$\exp\left(-\frac{ei}{\hbar c}\int_{P'}^{P''} A^\lambda \, \mathrm{d}x_\lambda\right)\cdot S_0$$

The quantity S will therefore contain this quantity as its most important term, which has the highest singularity on the light cone, whereby the integral again is to be taken along the straight line from P' to P''. We set

$$S = \exp\left(-\frac{ei}{\hbar c}\int_{P'}^{P''} A^\lambda \, \mathrm{d}x_\lambda\right)\cdot S_0 + S_1. \tag{12}$$

If one develops S_1 for small x_λ, then, according to (7), it must be representable in the form

$$S_1 = \frac{a}{x_\lambda x^\lambda} + b\log\left|\frac{x_\lambda x^\lambda}{C}\right| \tag{13}$$

In the last analysis, the density matrix is important only to calculate the charge-, current-, and energy-densities; thus it is sufficient (compare [eqn.] 2), to include only the terms up to third order in x_λ when developing the quantity a in x_λ, and to include only terms of first order in x^λ in the development of b; furthermore, for the same reason, it is sufficient to calculate only those terms which contain the α^λ only linearly. The following expressions for a and b are compatible with Equation (7) and with Dirac's results concerning the singularities of the density matrix (except for higher terms):

$$\left.\begin{aligned} a &= u\left\{\frac{ei}{24\hbar c}x_\rho x^\rho \alpha^\lambda\left(\frac{\partial F_{\lambda\sigma}}{\partial \xi_\rho} - \delta_\lambda^\rho \frac{\partial F_{\tau\sigma}}{\partial \xi_\tau}\right) - \frac{e^2}{48c^2\hbar^2}x_\rho x_\sigma x^\tau \alpha^\rho F^{\mu\sigma}F_{\mu\tau}\right\}, \\ b &= u\left\{\frac{ei}{24\hbar c}\alpha^\lambda \frac{\partial F_{\tau\lambda}}{\partial \xi_\tau} + \frac{e^2}{24\hbar^2 c^2}x_\lambda \alpha^\mu\left(F_{\tau\mu}F^{\tau\lambda} - \frac{1}{4}\delta_\mu^\lambda F_{\tau\sigma}F^{\tau\sigma}\right)\right\}. \end{aligned}\right\} \tag{14}$$

The field strengths are here always to be taken at the point $(x_1' + x_1'')/2 = \xi_1$. The quantity u is given by Equation (8).

If one defines S by Equations (12) through (14), the difference $R_S - S$ can still become singular on the light cone through the terms of the type $(x_\lambda x_\mu x_\nu x_\pi / x_\rho x^\rho) \cdot A^{\lambda\mu\nu\pi}$, or $x_\lambda x_\mu A^{\lambda\mu} \log|x_\rho x^\rho|$, or $(\alpha^\lambda \alpha^\mu - \alpha^\mu \alpha^\lambda) A_{\mu\lambda}/x_\rho x^\rho$. In these cases, however, one can deduce from the density matrix the current-, charge-, energy-, and momentum-densities, by making the limiting process $x_\lambda \to 0$ not on the light cone but from space-like or time-like directions. The above-mentioned singular terms then will make no contribution (the terms that are not linear in α_2 already drop out before the limiting process).

The matrices R, R_S, S, and S_0 are all Hermitian, i.e. they go over into their conjugates upon exchange of $x'k'$ with $x''k''$ (that is sign reversal of x).

Formula (14) is calculated most easily by the method given by Dirac (*loc. cit.*). The mathematical form of Expression (14) shows that the arbitrariness which still exists in specifying the quantities a and C, if no very complicated expressions for (14) are to be allowed, actually consists only in the fact that an expression of the form $x_\rho x^\rho \alpha_\lambda (\partial F_{\lambda\sigma}/\partial \xi_\sigma)$ and another one of the form $x_\rho x^\rho x_\lambda \alpha^\lambda F^{\tau\sigma} F_{\tau\sigma}$ could be added to a, without changing the singularities of the matrix S; furthermore, C is quite arbitrary. As regards the charge- and current-densities derived from the density matrix, these two cases of lack of definition (in a and C) similarly cause additive charge- and current-densities. However, the first term in a can be specified arbitrarily in the fashion given in (14), and all lack of definition in the charge density can be transferred to the quantity C. The second term in a, as will be shown in [eqn.] 2, is then determined by the conservation laws as given in Equation (14). The arbitrariness in the choice of the constant C finally is of no interest because, according to Dirac, the Equation $\mathcal{H}w = 0$ holds for the matrix w defined in (7); i.e. in the right-hand side of (11), the quantity C cancels out (except for terms which contain squares of x_λ or x^λ). However, this is true only if the electromagnetic field, together with all its derivatives, is continuous, and the matrix w can be developed in terms of x_λ and ξ_λ. If these assumptions are made, one has to accept the disadvantage that the theory cannot be simply connected with the special case of field-free space (e.g. by a perturbation calculation). If one allows discontinuous changes of higher derivatives of the fields or other singularities, the equation $\mathcal{H}w = 0$ no longer holds at the relevant singular points, and the specification of the quantity C becomes important. In this case, a suitable choice of the quantity C can be found by the following consideration: Let us consider a field originating from a prescribed external charge density, a field which is switched on adiabatically starting from field 'zero'. Then, this switch-on process will generate a matter field given by the matrix r; this matter field, as Equations (13) and (14) indicate, will wholly or partly compensate for or augment the external charge density; we now wish to choose C so that the total charge of the matter field given by r vanishes in the process under consideration; if this were not the case, the external charge density could not be separated from the generated electron charge density during the 'switch-on process', i.e. as the 'external' charge density one would already have defined the sums of the two densities. We shall return to the mathematical treatment of this question in [Section] 3. There we will also address the calculation of the quantity C – which, according to what we have said above, really has rather a mathematical than a physical significance; only its value will be given here:

$$C = 4\left(\frac{\hbar}{mc}\right)^2 \mathrm{e}^{-2/3-2\gamma}, \tag{15}$$

where γ represents the Euler constant: $\gamma = 0.577\ldots$.

We have thus specified the inhomogeneity in the differential Equation (11). As regards the currents that are derived from the density matrix r, our assumptions are equivalent to those of Dirac (*loc. cit.*); on the other hand, as Mr Dirac was kind enough to communicate, the present specifications of the matrix S yield a different energy and momentum density than the Dirac specification.

2 *The conservation laws*

From the density matrix r, the charge- and current-densities can be derived in the usual fashion. According to a study by Tetrode,[8] the energy- and momentum-tensors of the matter waves can be derived from the following equations:

$$\left.\begin{aligned} s_\lambda(\xi) &= -e\sum_{k'k''} \alpha^\lambda_{k'k''}(\xi k'|r|\xi k''); \\ U^\mu_\nu(\xi) &= \lim_{x\to 0}\left\{ic\hbar\frac{\partial}{\partial x_\mu} - \frac{e}{2}\left[A^\mu\left(\xi+\frac{x}{2}\right) + A^\mu\left(\xi-\frac{x}{2}\right)\right]\right\} \\ &\qquad\qquad \sum_{k'k''}\alpha^\nu_{k'k''}\left(\xi+\frac{x}{2}, k'|r|\xi-\frac{x}{2}, k''\right). \end{aligned}\right\} \tag{16}$$

To show that conservation laws of the usual form are valid for the quantities defined in this way, the following equation will first be proven:

$$\left.\begin{aligned} &\sum_{k'k''}\alpha^\lambda_{k'k''}\left[i\hbar\frac{\partial}{\partial\xi_\lambda} - \frac{e}{c}A^\lambda\left(\xi+\frac{x}{2}\right) + \frac{e}{c}A^\lambda\left(\xi-\frac{x}{2}\right)\right] \\ &\qquad\qquad \left(\xi+\frac{x}{2}, k'|r|\xi-\frac{x}{2}, k''\right) = 0 \end{aligned}\right\} \tag{17}$$

[8] H. Tetrode, *ZS. f. Phys.* **49**, 858, 1928.

except for terms which are at least quadratic in x_λ. Equation (17) is equivalent to the statement

$$\left.\begin{aligned}\sum_{k'k''}\alpha^\lambda_{k'k''}\left[i\hbar\frac{\partial}{\partial\xi_\lambda}-\frac{e}{c}A^\lambda\left(\xi+\frac{x}{2}\right)+\frac{e}{c}A^\lambda\left(\xi-\frac{x}{2}\right)\right]&\\ \left(\xi+\frac{x}{2},\,k'|S|\xi-\frac{x}{2},\,k''\right)=0&\end{aligned}\right\}\tag{18}$$

except for terms quadratic in x_λ; for the matrix R_S, the Equation $\mathcal{H}R_S=0$ is indeed valid, and thus surely Equation (17) is also. Now we have

$$\left.\begin{aligned}\left[i\hbar\frac{\partial}{\partial\xi_\lambda}-\frac{e}{c}A^\lambda\left(\xi+\frac{x}{2}\right)+\frac{e}{c}A^\lambda\left(\xi-\frac{x}{2}\right)\right]\exp\left(-\frac{ei}{\hbar c}\int_{P'}^{P''}A^\lambda\,\mathrm{d}x_\lambda\right)&\\ =\exp\left(-\frac{ei}{\hbar c}\int_{P'}^{P''}A^\lambda\,\mathrm{d}x_\lambda\right)\cdot\frac{e}{c}\int_{P'}^{P''}F^{\lambda\mu}\,\mathrm{d}x_\mu,&\end{aligned}\right\}\tag{19}$$

and $(\partial/\partial\xi_\lambda)S_0=0$. If we also consider that S_0 can be written in the form

$$\alpha^\lambda x_\lambda f(x^\rho x_\rho)+\beta mcg\,(x^\rho x_\rho)$$

it follows that Equation (18) in any case is correct for the first portion

$$\exp\left(-\frac{ei}{\hbar c}\int_{P'}^{P''}A^\lambda\,\mathrm{d}x_\lambda\right)\cdot S_0.$$

Now Equation (18) must thus still be demonstrated for the portion S_1. Its validity for the $b\log(x_\rho x^\rho/C)$ part of S_1 is here again obvious, because, according to Dirac, $\mathcal{H}w=0$ holds for the matrix (but compare p. 9 [pp. 173–4]). Thus there only remains the discussion of the $a/x_\lambda x^\lambda$ portion. The calculation shows that, according to (14), the terms originating from differentiation with respect to ξ_λ, in the first part of a, just cancel those in the second part, due to Equation (19). Thus the validity of Equation (17) has been proven.

If, in Equation (17), one proceeds to the limit $x_\lambda\to0$, there follows the charge conservation law:

$$\frac{\partial}{\partial\xi_\lambda}e\sum_{k'k''}\alpha^\lambda_{k'k''}(\xi k'|r|\xi k'')=\frac{\partial s_\lambda}{\partial\xi_\lambda}=0.\tag{20}$$

The limiting process $x_\lambda\to0$ is to be performed, according to the remarks in connection with Equations (14), not along the light cone but either from a space-like or time-like direction.

For the energy and momentum conservation law one finds similarly:

$$\begin{aligned}\frac{\partial U^\mu_\nu(\xi)}{\partial\xi_\nu}=&\lim_{x\to0}\left\{ic\hbar\frac{\partial}{\partial x_\mu}-\frac{e}{2}\left[A^\mu\left(\xi+\frac{x}{2}\right)+A^\mu\left(\xi-\frac{x}{2}\right)\right]\right\}\\ &\cdot\sum_{k'k''}\alpha^\nu_{k'k''}\frac{\partial}{\partial\xi_\nu}\left(\xi+\frac{x}{2},\,k'|r|\xi-\frac{x}{2},\,k''\right)\\ &-\lim_{x\to0}\frac{e}{2}\frac{\partial}{\partial\xi_\nu}\left[A^\mu\left(\xi+\frac{x}{2}\right)+A^\mu\left(\xi-\frac{x}{2}\right)\right]\\ &\cdot\sum_{k'k''}\alpha^\nu_{k'k''}\left(\xi+\frac{x}{2},\,k'|r|\xi-\frac{x}{2},\,k''\right),\end{aligned}$$

and, according to (17),

$$\frac{\partial U^\mu_\nu(\xi)}{\partial \xi_\nu} = \lim_{x\to 0}\left\{ic\hbar\frac{\partial}{\partial x_\mu} - \frac{e}{2}\left[A^\mu\left(\xi + \frac{x}{2}\right) + A^\mu\left(\xi - \frac{x}{2}\right)\right]\right\}$$

$$\left[-\frac{e}{ic\hbar}A^\nu\left(\xi + \frac{x}{2}\right) + \frac{e}{ic\hbar}A^\nu\left(\xi - \frac{x}{2}\right)\right]\sum_{k'k''}\alpha^\nu_{k'k''}\left(\xi + \frac{x}{2}, k'|r|\xi - \frac{x}{2}, k''\right)$$

$$- \lim_{x\to 0}\frac{e}{2}\frac{\partial}{\partial \xi_\nu}\left[A^\mu\left(\xi - \frac{x}{2}\right) + A^\mu\left(\xi - \frac{x}{2}\right)\right]$$

$$\times \sum_{k'k''}\alpha^\nu_{k'k''}\left(\xi + \frac{x}{2}, k'|r|\xi - \frac{x}{2}, k''\right)$$

$$= -eF^{\nu\mu}(\xi)\sum_{k'k''}\alpha^\nu_{k'k''}(\xi k'|r|\xi k'') = -F^{\nu\mu}s_\nu. \tag{21}$$

If one therefore adds to U^μ_ν the energy–momentum tensor of the Maxwell field:

$$V^\mu_r = \frac{1}{4\pi}\left(-F^{\tau\mu}F_{\tau\nu} + \frac{1}{4}\delta^\mu_\nu F^{\tau\sigma}F_{\tau\sigma}\right) \tag{22}$$

and if one uses as a basis the Maxwell equations in the form:

$$\frac{\partial F_{\tau\nu}}{\partial \xi_\nu} = -4\pi s_\tau \tag{23}$$

then for the tensor

$$T^\mu_\nu = U^\mu_\nu + V^\mu_\nu \tag{24}$$

the following relation holds:

$$\frac{\partial T^\mu_\nu}{\partial \xi_\nu} = 0. \tag{25}$$

According to Tetrode (*loc. cit.*), incidentally, the difference $U_{\mu\nu} - U_{\nu\mu}$ is a tensor whose divergence vanishes. The energy–momentum tensor of the matter field can thus also be symmetrized without disturbing the validity of (25).

The previous results can be summarized briefly in the following fashion: If one limits oneself to a correspondence-like intuitive theory of the matter field, the well-known difficulty of the appearance of negative energy levels in the Dirac theory can be avoided by replacing the homogeneous Dirac differential Equation (3) by an inhomogeneous equation, where the inhomogeneity is decisive for 'pair creation'. For the matter field satisfying this equation, the usual conservation laws are applicable together with the Maxwell field. At the same time, the energies of the matter field and those of the radiation field individually are always positive.

The invariance of the theory with respect to a sign change of the elementary charge can be recognized most simply in the following fashion: In Equations (11) and (16), let us replace $+e$ by $-e$ and furthermore $(x'k'|r|x''k'')$ by $-(x''k''|\bar{r}|x'k')$. The original Equations (11) and (16) then again hold for the matrix r.

3 *Applications*

Two simple examples shall illustrate the application of the methods described in [Sections] 1 and 2: First we assume that a scalar potential A_0, which can be regarded as a small perturbation, is slowly switched on and is then held constant, and we enquire into the matter created in the originally empty space; here, the charge density which gives rise to the potential A_0 shall be designated[9] as the 'external charge density'.

First we solve the Dirac differential equation for an electron, whose state before switching on the field is represented by a plane wave; let its eigenfunction be ψ_n, and let

$$\psi_n(x') = u_n(x') \exp\left(\frac{i}{\hbar}\overset{0}{p}_n x'_0\right) \tag{26}$$

hold before switching on the field. We set

$$\psi_n(x') = \sum_m c_{nm}(x'_0) u_m(x') \exp\left(\frac{i}{\hbar}\overset{0}{p}_m x'_0\right) \tag{27}$$

and from

$$\left\{\alpha^\lambda\left[i\hbar\frac{\partial}{\partial x'_\lambda} - \frac{e}{c}A^\lambda(x')\right] + \beta mc\right\}\psi = 0$$

there follows in the usual fashion:

$$\frac{d}{dx'_0}c_{nm} = \frac{i}{\hbar}H_{nm}\exp\left[\frac{i}{\hbar}\left(\overset{0}{p}_n - \overset{0}{p}_m\right)x'_0\right], \tag{28}$$

where

$$H_{nm} = \int u^*_m(x''')\frac{e}{c}\alpha^\lambda A^\lambda(x''')u_n(x''')\, dx'''. \tag{29}$$

Here, the integral dx''' means integration over the space variables and summation over the spin indices.

From (28) it follows that, if the H_{nm} have become constant in time:

$$c_{nm} = H_{nm}\frac{\exp\left[\frac{i}{\hbar}(\overset{0}{p}_n - \overset{0}{p}_m)x_0\right] - \varepsilon_{nm}}{\overset{0}{p}_n - \overset{0}{p}_m} + \delta_{nm}. \tag{30}$$

The constants ε_{nm} here depend on the type of time-rise of the H_{nm}; we shall assume that the rise occurs so slowly and uniformly that the ε_{nm} vanish to a sufficient approximation. Then one has

$$c_{nm} = H_{nm}\frac{\exp\left[\frac{i}{\hbar}(\overset{0}{p}_n - \overset{0}{p}_m)x_0\right]}{\overset{0}{p}_n - \overset{0}{p}_m} + \delta_{nm}$$

and

$$\psi_n(x') = \left[\sum_m u_m(x')\frac{H_{nm}}{\overset{0}{p}_n - \overset{0}{p}_m} + u_n(x')\right]\exp\left(\frac{i}{\hbar}\overset{0}{p}_n x'_0\right) \tag{31}$$

Terms of higher order than the first in H_{nm} will always be neglected in the following discussion. According to its definition, the following holds for the matrix R_S:

[9] This problem has essentially already been treated by Dirac in his Report for the Solvay Congress in 1933. [See Paper 3.]

$$(x'k'|R_S|x''k'') = \frac{1}{2}\left[\sum_{p_n^0>0} {}^n \psi_n^*(x'k')\psi_n(x''k'') - \sum_{p^0<0} {}^n \psi_n^*(x'k')\psi_n(x''k'')\right]. \tag{32}$$

The sum over all states can now be divided into an integral over the momenta and a sum over four possible states with each momentum. The operator

$$\frac{\alpha^l p^l + \beta mc}{|p^0|},$$

where in the numerator the summation over l runs only from 1 to 3 (as always with Latin indices), has the property that it yields $+1$ if it is applied to any state of positive energy, and -1 with a state of negative energy. By means of this operator, the summations over the spin states can thus be performed easily, and only the integrals over the momenta remain; in the following discussion, we shall always set $t' = t''$, i.e. $x^{0'} = x^{0''}$:

$$\begin{aligned}(x'k'|R_S|x''k'') = &-\frac{1}{2}\int\frac{\mathrm{d}\boldsymbol{p}}{h^3}\frac{\alpha^l p^l + \beta mc}{|p^0|}\exp\left[\frac{\mathrm{i}}{\hbar}p^\rho(x''_\rho - x'_\rho)\right]\\ &-\frac{1}{8}\int \mathrm{d}x'''\int\frac{\mathrm{d}\boldsymbol{p}'}{h^3}\int\frac{\mathrm{d}\boldsymbol{p}''}{h^3}\exp\frac{\mathrm{i}}{\hbar}[x'''_l(p'^l - p^{l''}) + p^{l''}x''_l - p^{l'}x'_l]\\ &\left\{\left(1 + \frac{\alpha^l p''^l + \beta mc}{|p^{0''}|}\right)\frac{\frac{e}{c}A^0(x''')}{|p^{0'}| + |p^{0''}|}\left(1 - \frac{\alpha^l p'^l + \beta mc}{|p^{0'}|}\right)\right.\\ &\left.+\left(1 - \frac{\alpha^l p''^l + \beta mc}{|p^{0''}|}\right)\frac{\frac{e}{c}A^0(x''')}{|p^{0'}| + |p^{0''}|}\left(1 + \frac{\alpha^l p'^l + \beta mc}{|p^{0'}|}\right)\right\}\\ &+ \text{conj.}\end{aligned} \tag{33}$$

The first term in (33) represents the matrix S_0 and is subtracted from R_S when forming r. The next two terms become

$$\begin{aligned}&-\frac{1}{4}\int \mathrm{d}x'''\int\frac{\mathrm{d}\boldsymbol{p}'}{h^3}\int\frac{\mathrm{d}\boldsymbol{p}''}{h^3}\exp\frac{\mathrm{i}}{\hbar}[x'''_l(p'^l - p''^l) + p''^l x''_l - p'^l x'_l]\\ &\cdot\frac{e}{c}\frac{A^0(x''')}{|p^{0\prime}| + |p^{0\prime\prime}|}\frac{|p^{0\prime}p^{0\prime\prime}| - p'_l p''_l - m^2c^2}{|p^{0\prime}p^{0\prime\prime}|}.\end{aligned} \tag{34}$$

To evaluate this expression, it is necessary to set:

$$\boldsymbol{p}' = \boldsymbol{f} + \frac{\boldsymbol{g}}{2};\ \boldsymbol{p}'' = \boldsymbol{f} - \frac{\boldsymbol{g}}{2};\ \boldsymbol{r}' - \boldsymbol{r}'' = \boldsymbol{r};\ \frac{\boldsymbol{r}' + \boldsymbol{r}''}{2} = \boldsymbol{R}. \tag{35}$$

One then has:

$$\begin{aligned}&-\frac{1}{4}\int \mathrm{d}x'''\int\frac{\mathrm{d}\boldsymbol{g}}{h^3}\frac{e}{c}A^0(\boldsymbol{r}''')\exp\left[\frac{\mathrm{i}}{\hbar}(\boldsymbol{r}''' - \boldsymbol{R})\boldsymbol{g}\right]\int\frac{\mathrm{d}\boldsymbol{f}}{h^3}\\ &\times\exp\left(-\frac{\mathrm{i}}{\hbar}\boldsymbol{fr}\right)\frac{|p^{0\prime}p^{0\prime\prime}| - \boldsymbol{p}'\boldsymbol{p}'' - m^2c^2}{(|p^{0\prime}| + |p^{0\prime\prime}|)|p^{0\prime}p^{0\prime\prime}|}.\end{aligned} \tag{36}$$

The fraction under the integral sign is best developed in terms of $\boldsymbol{g}$ for $g \ll mc$, and one obtains the value

$$\frac{1}{2k_0^3}\left[\frac{g^2}{2} - \frac{(\boldsymbol{fg})^2}{2k_0^2} - \frac{3g^4}{16k_0^2} + \frac{5(\boldsymbol{fg})^2 g^2}{8k_0^4} - \frac{7(\boldsymbol{fg})^4}{16k_0^6} + \cdots\right], \tag{37}$$

where we have set $k_0^2 = k^2 + m^2c^2$. A longer calculation for (36) leads to the result (for small values of $|\boldsymbol{r}| = r$):

$$-\frac{\pi}{4}\int \mathrm{d}x''' \int \frac{\mathrm{d}\boldsymbol{g}}{h^3}\frac{e}{c}A^0(\boldsymbol{r}''')\exp\left[\frac{\mathrm{i}}{\hbar}(\boldsymbol{r}''' - \boldsymbol{R})\boldsymbol{g}\right]$$

$$\cdot\frac{1}{h^3}\left[g^2\left(\frac{2}{9} - \frac{2}{3}\gamma - \frac{2}{3}\log\frac{mcr}{\hbar}\right) + \frac{1}{3}\frac{(\boldsymbol{gr})^2}{r^2} - \frac{g^4}{15m^2c^2}\right]$$

$$= \frac{1}{16\pi h}\left[\frac{2}{3}\left(\frac{1}{3} - \gamma - \log\frac{mcr}{\hbar}\right)(\mathrm{grad}_{\boldsymbol{R}})^2 + \frac{1}{3r^2}(\boldsymbol{r}\,\mathrm{grad}_{\boldsymbol{R}})^2\right.$$

$$\left. + \frac{1}{15}\left(\frac{\hbar}{mc}\right)^2(\mathrm{grad}_{\boldsymbol{R}})^2(\mathrm{grad}_{\boldsymbol{R}})^2\right]\frac{e}{c}A^0(\boldsymbol{R}). \tag{38}$$

The first two terms – after they have been doubled, since the complex conjugate must still be added to (36) – represent the portions

$$\frac{a}{x_\lambda x^\lambda} + b\log\left|\frac{x_\lambda x^\lambda}{C}\right|$$

of Equation (13) and therefore should be omitted, when one goes over from R_S to the matrix r. [Eq. (38) 'contains a computational error which led to a different value for C (see equation following Eq. (38)); there the letter γ, in contradiction to general usage, was also used for the logarithm of the Euler constant'.] Formula (38) also substantiates retrospectively the fact that the constant C in Equation (15) was set equal to $4(\hbar/mc)^2\exp(-2/3 - 2\gamma)$. In this way we reach the result that it becomes unnecessary to correct the field-generating total charge with each new step of the perturbation calculation. Finally, the density matrix $(x'k'|r|x''k'')$ for $|\boldsymbol{r}| = 0$ becomes

$$(\xi k'|r|\xi k'') = \frac{1}{120\pi h}\frac{e}{c}\left(\frac{\hbar}{mc}\right)^2 \Delta\Delta A^0(\xi) \tag{39}$$

and the charge density itself [$\Delta A^0(\xi) = -4\pi\rho_0$, where ρ_0 designates the external charge density]:

$$\rho = -\frac{1}{15\pi}\frac{e^2}{\hbar c}\left(\frac{\hbar}{mc}\right)^2\Delta\rho_0, \tag{40}$$

as has already been calculated by Dirac[10]. This additional density, whose total charge vanishes, also has no physical significance; for it cannot be separated from the 'external' density and is therefore automatically accounted to the 'external' density.

The 'polarization of the vacuum' becomes a physical problem only with time-varying external densities; for example, let us consider a charge distribution which is moved

[10] P. A. M. Dirac, Report for the Solvay Congress 1933; the Dirac value differs from the above by a factor of 2. As Mr Dirac was kind enough to communicate to me, this came from an oversight in his equations.

periodically back and forth. In such a case, the external charge density can be divided into its time average and into a second density which fluctuates periodically about the value zero. The space integral of the second part vanishes, if the external charge density is moved back and forth in a finite spatial region. For the first part, the previous considerations apply, and for this part the 'polarization of the vacuum' plays no physical role. The total charge of a particle can thus never be changed by the polarization of the vacuum. In order to see what happens with the second part, we consider in Equations (26) through (29), in place of the time-constant scalar potential A^0, a potential which varies periodically, and we set

$$A^0(x') = B^0(\boldsymbol{r}')\exp\left(\frac{\mathrm{i}}{\hbar}fx'_0\right) + \text{conj.} \tag{41}$$

The only change which must then be made in Expressions (34) through (36) consists in replacing the fraction

$$\frac{1}{(|p^{0'}| + |p^{0''}|)|p'_0 p''_0|}$$

by

$$\frac{|p^{0'}| + |p^{0''}|}{[(|p^{0'}| + |p^{0''}|)^2 - f^2]|p^{0'}p^{0''}|}.$$

The new formulas are simply derived from the old ones – we assume $f \ll mc$ – by multiplying the expression under the integral sign in (36) by $1 + f^2/4k_0^2$. Furthermore, terms with α^1 occur in the density matrix, however; but we shall limit ourselves to calculating the charge density, for which the terms of α^1 play no role. If we consider only terms proportional to g^2 in (37), the expression

$$\frac{1}{2k_0^3}\left[\frac{g^2}{2} - \frac{(fg)^2}{2k_0^2}\right]\frac{f^2}{4k_0^2} \tag{42}$$

is newly added to (37). The corresponding part of the density matrix thus becomes

$$-\frac{\pi}{4\cdot 15}\int \mathrm{d}x'''\int\frac{\mathrm{d}g}{h^3}\frac{e}{c}\left[B^0(\boldsymbol{r}''')\exp\left(\frac{\mathrm{i}}{\hbar}fx_0'''\right) + \text{conj.}\right]$$
$$\times \exp\left[\frac{\mathrm{i}}{\hbar}(\boldsymbol{r}''' - \boldsymbol{R})\boldsymbol{g}\right]\frac{f^2g^2}{h^3(mc)^2} \tag{43}$$

and consequently the additional density is

$$\rho = -\frac{1}{15\pi}\frac{e^2}{\hbar c}\frac{f^2}{(mc)^2}\cdot\rho_0 \tag{44}$$

Here, ρ_0 designates the periodically fluctuating density, which causes the field $B^0(x')\exp[(\mathrm{i}/\hbar)fx'_0]$, and whose space integral vanishes. Equation (44) shows that the dipole moment associated with an oscillating charge is reduced by the polarization of the vacuum and that this effect increases the higher the frequency of oscillation. As was already emphasized by Dirac, this should entail a modification of the Klein–Nishina scattering formula. However, in the region of the Compton wavelength, this will only amount to about one-tenth of one-tenth of one per cent.

If one performs an analogous calculation so as to calculate the matter density induced by a light wave, the result is that the periodically alternating field of a monochromatic plane light wave generates neither a charge density nor a current density. One can easily see that this result also remains correct in an arbitrary approximation: By an electromagnetic field in empty space, [we mean that] one cannot distinguish the sign of the charge; thus the induced charge density must vanish. For invariance reasons, the current density then also vanishes. Of course, the vanishing of the energy density does not yet follow from this, and in fact two mutually penetrating plane light waves can cause the creation of matter. However, the intuitive theory of matter waves is no longer adequate for treating such problems (pair creation and annihilation), and we shall therefore proceed to the quantum theory of waves.

II Quantum theory of wave fields

1 Setting up the basic equations

In the quantum theory of matter waves, the product of the wave function with its conjugate corresponds to the Dirac density matrix; we therefore set

$$R = \psi^*(x'k')\psi(x''k''). \tag{45}$$

The following commutation relation holds for the wave function (for $x_0' = x_0''$):

$$\psi^*(x'k')\psi(x''k'') + \psi(x''k'')\psi^*(x'k') = \delta(x'x'')\delta_{k'k''}. \tag{46}$$

If one regards the Maxwell field as a given c-number field, the Dirac density matrix is simply the expectation value of the matrix defined by (45). Because of the commutation relation (46), the following holds in the quantum theory of waves:

$$R_S = \tfrac{1}{2}[\psi^*(x'k')\psi(x''k'') - \psi(x''k'')\psi^*(x'k')]. \tag{47}$$

The equations

$$\mathcal{H} R_S = 0 \tag{3a}$$

and $R_S = r + S$ remain unchanged, and a change due to the non-commutability of the field strengths with the potentials could become necessary only in the form of the inhomogeneity $\mathcal{H}S$ in

$$\mathcal{H}r = -\mathcal{H}S \tag{11a}$$

Now, no non-commuting functions occur in the first term

$$\exp\left(\frac{ei}{\hbar c}\int_{P'}^{P''} A^\lambda \,\mathrm{d}x_\lambda\right)\cdot S_0$$

Terms occur in S_1 (compare (13) and (14)), which are quadratic in the field strength and which play a role if one calculates the energy and momentum densities from the density matrix. As long as one limits oneself to the calculation of charge and current densities, these terms do not appear. Now, since Maxwell's equations together with the inhomogeneous Equation (11a) completely determine the physical process, the formalism described in [Part] I can be transferred into quantum mechanics by a procedure which was given for ordinary quantum electrodynamics in a note by the author[11] in connection with a previous study by Klein[12]. This procedure starts from Maxwell's

[11] W. Heisenberg, *Ann. d. Phys.* **9**, 338, 1931.
[12] O. Klein, *ZS. f. Phys.* **41**, 407, 1927.

equations and the wave equation, which are treated as q-number relations and which are integrated by the usual methods of the intuitive theory. Ordinarily, a perturbation method is used in the integration of the basic equations. Here, the interaction between light and matter is assumed to be small, and one expands in powers of the charge. The plane light waves in empty space and the plane electron waves in a field-free space then appear as the unperturbed system. Such a perturbation method can also be used readily in the present theory. For this, it is also necessary to expand the matrix S, which determines the inhomogeneity of the wave equation, in powers of the charge, and to take into account the individual terms of the expansion in the perturbation method in sequence according to various approximations. In the zeroth approximation, therefore, one will merely have to subtract the matrix S_0 from R_S in order to proceed from R_S to r and thus to the charge and current densities. If the wave function is represented in the form

$$\psi(xk) = \sum_n a_n u_n(xk), \tag{48}$$

where the equations

$$a_n a_m^* + a_m^* a_n = \delta_{nm} \tag{49}$$

then hold, one obtains (below we shall always set $x_0' = x_0''$)

$$\begin{aligned} R_S &= \tfrac{1}{2}[\psi^*(x'k')\psi(x''k'') - \psi(x''k'')\psi^*(x'k')] \\ &= \sum_{n,m} \frac{1}{2}(a_n^* a_m - a_m a_n^*) u_n^*(x'k') u_m(x''k''). \end{aligned} \tag{50}$$

From this follows for r, taking into account the definition of S_0:

$$r = \sum_{n,m} \frac{1}{2}\left(a_n^* a_m - a_m a_n^* + \frac{p_n^0}{|p_n^0|} \cdot \delta_{nm}\right) u_n^*(x'k') u_m(x''k''). \tag{51}$$

According to Jordan and Wigner[13], the operators a_n are represented in the form

$$a_n^* = N_n \Delta_n V_n; \; a_n = V_n \Delta_n N_n, \tag{52}$$

where Δ_n transforms the number N_n into $1 - N_n$, and where one sets

$$V_n = \prod_{t \leqslant n} (1 - 2N_t).$$

For states of negative energy, one can now introduce[14]:

$$\left.\begin{aligned} a_n^* &= a_n' = V_n' \Delta_n' N_n' = V_n \Delta_n N_n', \\ a_n &= a_n'^* = N_n' \Delta_n' V_n' = N_n' \Delta_n V_n \end{aligned}\right\} \tag{53}$$

One then has $N_n' = 1 - N_n$.

For the matrix r one finally obtains:

$$\begin{aligned} r = &\sum_{p_{0n}>0} a_n^* a_n u_n^*(x'k') u_n(x''k'') - \sum_{p_{0n}<0} a_n'^* a_n' u_n^*(x'k') u_n(x''k'') \\ &+ \sum_{n \neq m} a_n^* a_m u_n^*(x'k') u_m(x''k'') \end{aligned}$$

[13] P. Jordan and E. Wigner, *ZS. f. Phys.* **47**, 631, 1928.
[14] Compare W. Heisenberg, *Ann. d. Phys.* **10**, 888, 1931.

$$= \sum_{p_{0n}>0} {}_nN_n u_n^*(x'k')u_n(x''k'') - \sum_{p_{0n}<0} {}_nN'_n u_n^*(x'k')u_n(x''k'')$$

$$+ \sum_{n\neq m} a_n^* a_m u_n^*(x'k')u_m(x''k''). \tag{54}$$

This representation of the density matrix agrees with the representations which were chosen by Pauli and Peierls,[15] Oppenheimer and Furry, and Fock (*loc. cit.*). N_n denotes the number of electrons, N'_n the number of positrons, and the symmetry of the theory in the sign of the charge is guaranteed in advance. However, this representation is correct only in the zeroth approximation. If one goes to the first approximation, on the one hand the coefficients a_n as functions of time will also contain terms which are linear in the external field strengths (compare e.g. W. Heisenberg, *Ann. d. Phys.* 9, 341, Equation (9) [see Paper 2]), and on the other hand, to form r, the terms of the matrix S which are linear in e will have to be subtracted, that is the terms

$$-\frac{ei}{\hbar c}\int_{P'}^{P''} A^\lambda \, dx_\lambda \cdot S_0 + \frac{e}{48\pi^2\hbar c}\left\{\frac{x_\rho x^\sigma}{x_\tau x^\tau}\cdot\alpha^\lambda\left(\frac{\partial F_{\lambda\sigma}}{\partial\xi_\rho} - \delta_\lambda^\rho\frac{\partial F_{\mu\sigma}}{\partial\xi_\mu}\right)\right.$$

$$\left. + \alpha^\lambda\frac{\partial F_{\tau\lambda}}{\partial\xi_\tau}\log\left|\frac{x_\rho x^\rho}{C}\right|\right\}. \tag{55}$$

These terms, together with the terms in the coefficients a_n, which are linear in e, then yield an addition to the matrix r which leads to finite charge and current densities (first approximation), and which can be used to calculate the electromagnetic fields in the second approximation, etc.

Instead of this procedure, which is closely connected with the integration method of the intuitive theory, one can also form a Hamiltonian function in the usual fashion, and then implement the perturbation theory in the associated Schrödinger equation. For this purpose, we use the expression for the total energy which follows from Equation (16), but do not, however, pass to the limit $x^\lambda = 0$. The total energy then takes the form

$$E = \int d\xi\left\{-\left(ci\hbar\frac{\partial}{\partial x_l} - \frac{e}{2}\left[A^l\left(\xi + \frac{x}{2}\right)\right.\right.\right.$$

$$\left.\left. + A^l\left(\xi - \frac{x}{2}\right)\right]\right)\sum_{k'k''}\alpha^l_{k'k''}\sum_{nm}\frac{1}{2}(a_n^*a_m - a_m a_n^*)u_n^*\left(\xi + \frac{x}{2}, k'\right)u_m\left(\xi - \frac{x}{2}, k''\right)$$

$$- \sum_{k'k''}\beta_{k'k''}mc^2\sum_{nm}\frac{1}{2}(a_n^*a_m - a_m a_n^*)u_n^*\left(\xi + \frac{x}{2}, k'\right)u_m\left(\xi - \frac{x}{2}, k''\right)$$

$$-\sum_{k'}\left(ci\hbar\frac{\partial}{\partial x_0} - \frac{e}{2}\left[A^0\left(\xi + \frac{x}{2}\right) + A^0\left(\xi - \frac{x}{2}\right)\right]\right)\left(\xi + \frac{x}{2}, k'|S|\xi - \frac{x}{2}, k'\right)$$

$$\left. + \frac{1}{8\pi}(\boldsymbol{E}^2 + \boldsymbol{H}^2)\right\}. \tag{56}$$

If one expands the Hamiltonian function again in powers of the elementary charge, and if one further deletes the zero-point energy of the radiation in the terms $\boldsymbol{E}^2 + \boldsymbol{H}^2$ as well as in the corresponding terms of Expression (14), one obtains, in the limit $x \to 0$, for the Hamiltonian function of zeroth order:

[15] I sincerely thank Mr W. Pauli for communicating these results in a letter.

$$H_0 = \sum_{E_n>0} N_n E_n - \sum_{E_n<0} N'_n E_n + \sum_{ge} M_{ge} h\nu_{ge}, \tag{57}$$

where we have set $E_n = -cp^0_n$, and M_{ge} is the number of light quanta in the state $\boldsymbol{g}$ with the polarization $\boldsymbol{e}$. For the perturbation energy of the first order, in the limit $x \to 0$ (A^0 was here set to 0 for the sake of simplicity), one likewise obtains:

$$H_1 = \int \mathrm{d}\xi eA^l(\xi) \sum_{k'k''} \alpha^l_{k'k''} \left[\sum_{E_n>0} N_n u^*_n(\xi k') u_n(\xi k'') - \sum_{E_n<0} N'_n u^*_n(\xi k') u_n(\xi k'') + \frac{1}{2} \sum_{n \neq m} (a^*_n a_m - a_m a^*_n) u^*_n(\xi k') u_m(\xi k'') \right]. \tag{58}$$

In the expressions for H_0 and H_1, the present theory therefore agrees with the results of Oppenheimer and Furry, Peierls, and Fock. However, we also obtain terms of higher order, which come from the matrix S. In these terms, one cannot immediately make a transition to the limit $x \to 0$. Rather, when performing the perturbation calculation up to a second approximation, the terms in H_2 must first be combined with the terms of the type $H_1^{nl} H_1^{lr}/W_n - W_l$ which come from H_1. Only then can the limiting process $x \to 0$ be performed, and this yields a definite result for the second-order energy.

In this fashion, the perturbation method can be continued in principle, unless an infinite self-energy causes the method to diverge, as in the previous quantum electrodynamics[16]. The perturbation energy H_2 has the following form:

$$H_2 = \int \mathrm{d}\xi \left[\mathrm{i}\frac{e^2}{\hbar c} \left(\int A^\lambda \, \mathrm{d}x_\lambda \right)^2 \frac{\partial}{\partial x_0} S_0 + \frac{1}{48\pi^2} \frac{e^2}{\hbar c} \frac{x_\lambda x^\sigma}{x_\rho x^\rho} A^\lambda \left[\frac{\partial F_{0\sigma}}{\partial \xi_0} - \frac{\partial F_{\tau\sigma}}{\partial \xi_\tau} \right] - \frac{1}{96\pi^2} \frac{e^2}{\hbar c} \frac{x_\sigma x^\tau}{x_\rho x^\rho} F^{\mu\sigma} F_{\mu\tau} + \frac{1}{48\pi^2} \frac{e^2}{\hbar c} \log \left| \frac{x_\rho x^\rho}{C} \right| \cdot \left(F_{\tau 0} F^{\tau 0} - \frac{1}{2} F_{\tau\mu} F^{\tau\mu} \right) \right]. \tag{59}$$

Because of the integration over ξ, H_2 gives rise only to matrix elements which correspond to the creation or annihilation of light quanta with the same momentum; for ordinary processes, where light quanta are emitted or absorbed or scattered, these matrix elements therefore play no role in a first approximation. The perturbation energy H_3 has the form

$$H_3 = \int \mathrm{d}\xi \frac{e^3}{6c^2\hbar^2} (A^\lambda x_\lambda)^3 \frac{\partial S_0}{\partial x_0}. \tag{60}$$

Here, three light quanta whose momentum sums to zero are combined. Finally H_4 is reduced to the term

$$H_4 = \int \mathrm{d}\xi \left[-\mathrm{i}c\hbar \frac{1}{24} \left(-\frac{\mathrm{i}e}{\hbar c} A^\lambda x_\lambda \right)^4 \frac{\partial S_0}{\partial x_0} \right] = \frac{1}{48\pi^2} \left(\frac{e^2}{\hbar c} \right)^2 \frac{1}{\hbar c} \int \mathrm{d}\xi \frac{(A^\lambda x_\lambda)^4}{(x_\rho x^\rho)^2} \tag{61}$$

[16] In this connection compare V. Weisskopf, *ZS. f. Phys.* **89**, 27, 1934: furthermore also the attempt to avoid the infinite self-energy of the electron by M. Born, *Proc. Roy. Soc. London (A)* **143**, 410, 1934; M. Born and L. Infeld, *ibid.* **144** 425, 1934.

and gives rise to matrix elements which lead to the scattering of light from light (annihilation and creation of two light quanta each, with the same momentum sum). Halpern[17] and Debye[18] have already independently pointed out the fact that the Dirac theory of the positron predicts the scattering of light from light – even where the energy of the light quanta is not sufficient for pair creation. The matrix elements in H_4, however, provide no information concerning the magnitude of this scattering, since they must first be combined with contributions originating from lower approximations, in order to yield a measure of the probability of a scattering process. Higher perturbation elements than H_4 do not occur; H_5, H_6, etc. all vanish in the limit $x = 0$.

2 *Applications*

For most practical applications, e.g. pair creation, annihilation, Compton scattering, etc., the theory presented here does not yield anything new compared to previous formulations of the Dirac theory. For, in all the above cases, the perturbation calculation can be terminated with the second approximation, and the new terms in H_2, because of their special form, contribute nothing to the desired transition probabilities. It is different with the above-mentioned problem of the scattering of light from light and with the coherent scattering of γ-rays at fixed charge centers, which has been discussed by Delbrück.[19] The calculation of these problems, however, is so complicated that it will not be attempted here.

We therefore shall restrict the applications to an example where the terms H_2 in Equation (59) become important. We shall treat the matter density associated with a light quantum and in particular the self-energy of the light quantum derived on the basis of this matter density. If one first neglects the terms H_2 and performs calculations using previously acceptable methods, the process presents itself as follows: Since matrix elements occur in H_1 (Equation (58)) which correspond to the transformation of a light quantum into a pair, a light quantum will create a matter field in its neighborhood – similar to an electron creating a Maxwell field in its neighborhood. The energy of this matter field becomes infinite, in precise analogy to the infinite self-energy of the electrons. A portion of the singular terms in the infinite self-energy of the light quanta now vanishes if one takes into account the perturbation elements H_2. For these have precisely such a character that, for a classical light wave, no infinite self-energy would occur. Nevertheless, the following calculation shows that an infinite portion of the self-energy remains, a portion which is caused by the application of quantum theory. The analogy with the self-energy of the electrons is here complete; for even a continuous charge distribution would lead to a finite self-energy according to the Maxwell theory; only the 'quantization' of the charge distribution leads to an infinite self-energy. If one represents the quantization of the electromagnetic field by the picture of point-like light quanta, the process by which the self-energy becomes infinite is also illuminating in the intuitive theory of matter waves, since the inhomogeneity in Equation (11) contains the field strengths and its first and second derivatives, which become singular in the neighborhood of the light quantum.

17 O. Halpern, *Phys. Rev.* **44**, 885, 1934.
18 I would like to thank Mr Debye sincerely for kindly communicating to me his ideas.
19 M. Delbrück, discussion of experimental results of Miss L. Meitner and her colleagues, *ZS. f. Phys.* **84**, 144, 1933.

To calculate the desired self-energy, one can start from a well-known formula of perturbation theory for the second-order energy:

$$W_2 = H_0 s_1^2 - s_1 H_0 s_1 + s_1 H_1 - H_1 s_1 + H_2. \tag{62}$$

Here, H_0, H_1, H_2 designate the different terms of the Hamiltonian function, and s_1 is the first term of the characteristic matrix s for the canonical transformation

$$W = sHs^{-1}, \tag{63}$$

where

$$s = 1 + s_1 + \cdots. \tag{64}$$

Following the sense of the method described in the previous section, the matrices H in Equation (62) should first be taken for finite distances x_λ; only at the end should one pass to the limit $x_\lambda \to 0$. The matrix s_1 can be calculated from the value of H_1 in the limit $x_\lambda = 0$ in the usual fashion:

$$s_1^{lm} = \frac{H_{1_{(x=0)}}^{lm}}{W_0^l - W_0^m}. \tag{65}$$

The element of the matrix $H_1(x = 0)$, which belongs to the simultaneous creation of an electron with momentum $\boldsymbol{p}''$, a positron of momentum $\boldsymbol{p}'$, and the annihilation of a light quantum of momentum $\boldsymbol{g}$ (polarization direction $\boldsymbol{e}$), has the form:

$$\frac{e\hbar}{\sqrt{V}}\left(\frac{c}{g}\right)^{1/2}(\boldsymbol{p}', p_0 < 0|\alpha \boldsymbol{e}|\boldsymbol{p}'', p_0'' > 0) \cdot M_{g,e}^{1/2}, \tag{66}$$

where V represents the volume which is given by the periodic boundary conditions, and $M_{g,e}$ represents the number of light quanta in the state $\boldsymbol{g}$, $\boldsymbol{e}$. Furthermore, we set

$$(\boldsymbol{p}', p_0' < 0|\alpha \boldsymbol{e}|\boldsymbol{p}'', p_0'' > 0) = \int \mathrm{d}\boldsymbol{r} \sum_{k'k''} u^*_{\boldsymbol{p}', p_0' < 0}(\alpha^l e_l) u_{\boldsymbol{p}'', p_0'' > 0} \tag{67}$$

We now insert into Equation (62) the expressions for s_1 which follow from (65) and (66), where we must consider not only the matrix elements (66) but also those corresponding to the process: simultaneous creation of electron, positron, and light quantum. Then we obtain contributions from the terms $s_1 H_1 - H_1 s_1$, which remain finite as long as $x_\lambda x^\lambda$ does not vanish and which, when they are combined with the corresponding terms in H_2, yield a finite contribution to W_2 even in the limit $x_\lambda \to 0$. However, this does not hold for the portions $(H_0 s_1 - s_1 H_0)s_1$. If H_0 is decomposed into a part belonging to the matter waves and a part belonging to the light waves, the first part also yields a contribution that is finite for $x_\lambda x^\lambda \neq 0$, and this part, when combined with H_2, makes a finite contribution to W_2 in the limit $x_\lambda \to 0$. The part associated with the electromagnetic field, however, does not depend at all on x_λ; it leads to the sum

$$\sum_{\boldsymbol{p}'+\boldsymbol{p}''=\boldsymbol{g}} |(\boldsymbol{p}'|s_1|\boldsymbol{p}'')|^2 gc. \tag{68}$$

This sum diverges; Expression (68) can immediately be designated as the infinite self-energy of the light quanta. If the summation in (68) is carried only to large but finite values $|\boldsymbol{p}'| = P$, and if only that part in (62) is considered that is proportional to $M_{g,e}$, an expression is obtained of the form

$$g \cdot c \cdot M_{g,e} \cdot \frac{e^2}{\hbar c} \log \frac{P}{mc}. \tag{69}$$

In the quantum theory of wave fields, the application range of the Dirac formulation of positron theory therefore is not significantly greater than the application range of the elementary formulas of Pauli, Peierls, Fock, Oppenheimer and Furry. Equations (48) through (61), however, show how these formulas can be regarded as the first steps of a consistent approximation method which satisfies the requirements of relativistic- and gauge-invariance; furthermore, the formalism described here yields finite expectation values for the current- and energy-densities in a first approximation, even where the elementary formulas lead to infinite values. That divergences would appear already in the second approximation of the quantum theory of wave fields was expected according to the previous results of quantum electrodynamics.

The fact that only the application of quantum theory leads to divergences which do not appear in the intuitive theory of wave fields, suggests the conjecture that this intuitive theory already does indeed contain the correct correspondence-like description of the process, but that the transition to quantum theory cannot be made as primitively as is being attempted in presently available theories. In the Dirac theory of the positron, a clean separation of the fields which actually occur, into matter fields and electromagnetic fields, is scarcely possible any more; this appears especially from the fact that, in the quantum theory of waves, the matrix R_S – and not the matrix r – can be represented simply by the matter wave functions ψ. Only in a unified theory of matter and light fields, which would assign a definite value to the Sommerfeld constant $e^2/\hbar c$, would a consistent combination of the requirements of quantum theory and those of correspondence with the intuitive field theory be possible.

7 The quantization of the scalar relativistic wave equation

W. PAULI and V. WEISSKOPF

Helvetica Physica Acta, 7: 709–31 (1934). Received 27 July 1934.

Summary. The present paper consistently applies the Heisenberg–Pauli formalism of the quantization of wave fields to the scalar relativistic wave equation for matter fields when the particles have Bose–Einstein statistics. Without any further hypothesis, this yields the existence of particles with opposite charge but with the same rest mass, which can be created or destroyed with the absorption or respectively emission of electromagnetic radiation. The frequency of these processes proves to be of the same order of magnitude as the frequency for particles of the same charge and mass, which follows from the Dirac hole theory (§ 4). Here we investigate the possibility of charges with opposite charge, without spin, and with Bose–Einstein statistics, conformable to the correspondence principle and likewise satisfying relativity requirements. Compared to the hole theory, this possibility has the advantage that the energy is by itself always positive. However, just as in the original version of the hole theory, the theory discussed here leads not only to infinitely large self-energies but also to an infinite polarizability of the vacuum.

§1 The connection of the scalar relativistic wave equation with the existence of oppositely charged particles

By introducing the operators

$$E = \mathrm{i}h\frac{\partial}{\partial t},\ p_k = -\,\mathrm{i}h\frac{\partial}{\partial x^k} \qquad (k = 1, 2, 3) \tag{1}$$

the scalar relativistic wave equation in the force-free case can be written

$$\frac{E^2}{c^2} - \sum_{k=1}^{3} p_k^2 - m^2c^2 = 0 \tag{2}$$

(Here and below h is always Planck's constant divided by 2π, m is the rest mass of the electron, and c is the speed of light.) As is well known, this wave equation has generally been relinquished in favor of the Dirac four-component wave equation since the former does not yield the spin of the particles and thus surely represents for electrons an inadequate approximation to experiment. A special justification is therefore required if below we again resume the discussion of the consequences of the first wave equation. The empirical discovery of the positron was theoretically interpreted by Dirac in his re-interpretation of the negative energy states which occurred in his original theory. We show that this requires a revision of Dirac's *a priori* arguments, which are based on general quantum-mechanical transformation theory. These arguments originally substantiated his transition from the scalar relativistic wave equation to his spinor wave equation. By showing that this revision is necessary, we believe we have given the justification mentioned above. Indeed, we shall show below that, by applying to the present problem the general methods for quantizing wave fields, which previously were formulated by Heisenberg and Pauli[1], not only can no general objections be maintained against the scalar wave equation from the standpoint of quantum-mechanical transformation theory, but also relativistic and gauge invariance of the theory can be preserved, and, *without any further hypothesis, one arrives at the conclusion that oppositely charged particles exist and that processes occur in which such particle pairs are created and annihilated, and where furthermore the energy of the matter-wave field automatically always comes out positive*. For the particles, one must here assume the statistics of symmetric states (Bose–Einstein statistics), but surely it is only to be regarded as satisfactory that, without simultaneously introducing the spin, the exclusion principle cannot be introduced while preserving the relativistic invariance of the theory.

Concerning the above-mentioned *a priori* argument of Dirac against the scalar relativistic wave equation[2], this is essentially based on two presuppositions.

1. In relativistic quantum theory, it is possible to formulate a *single-body problem* self-consistently.
2. The spatial particle density $\rho(x)$ (which is to be interpreted statistically) is a meaningful concept. After integration over an arbitrary finite volume, one obtains from it an 'observable' (in the sense of transformation theory) with the eigenvalues 0 and +1.

As soon as the first presupposition applies, it is indeed unnecessary to apply to the problem the formalism of the quantization of wave fields; rather, it is then possible to make do with the ordinary wave field in three-dimensional space. The consequence of the second presupposition is that the particle density not only must be the fourth component of a four-vector and must satisfy a continuity equation, but that it must also have the property of never being negative. Furthermore, the eigenvalues of the associated density matrix, after integration over an infinite volume, become the correct ones, as Dirac had shown, only if the particle density has the form:[3]

$$\rho(x) = \sum_r \psi_r^* \psi_r .$$

On the other hand, the particle density which is associated with the scalar relativistic wave equation has the form

$$\rho(x) = \psi^* \left(ih \frac{\partial \psi}{\partial t} - e\Phi_0 \psi \right) - \left(ih \frac{\partial \psi^*}{\partial t} + e\Phi_0 \psi^* \right) \psi \tag{3}$$

where e signifies the charge of the particle and Φ_0 is the external scalar potential. Since this does not have the required form, a contradiction seems to be established.

[1] *Zeitschr. f. Phys.* **56**, 1, 1929.

[2] This is presented in most detail in the Leipzig Lectures 1932 (appeared in collective form under the title *Quantum Theory and Chemistry*), p. 85 ff.

[3] Only from this form for ρ does one further conclude that the wave equations must be of first order in $\partial/\partial t$.

As is well known, Dirac – supported by the fact that, on the basis of his wave equation, a wave packet consisting of negative energy states moves in an external field as would correspond to a particle with opposite charge, the same mass, and positive energy – has drawn upon the states of negative energy to interpret the positron in the following fashion. Supposedly only the deviations from the case where all states of negative energy are occupied, the 'holes' in the occupation of the negative energy states, are observable, that is contribute to the 'true' (field producing) charge density and to the actual (then positive) energy.

The difficulties of a self-consistent and relativistically and gauge invariant formulation of this Dirac hole theory of electrons and positrons in the case of external fields have indeed been discussed several times in the literature. Without dealing with these difficulties in more detail, we can make the following observations:

1. Because of the processes of pair creation, and because of the new interpretation of the negative energy states altogether, it is no longer possible to limit oneself to a single-body problem.
2. The particle density no longer has a direct physical meaning.[4] However, in the force-free case, the number of particles with a given momentum (probability density in momentum space), and consequently also a total number of particles present, are meaningful 'observables'.
3. On the other hand, not only the total charge but also the charge density $\rho(x)$ is a meaningful observable. After integration over an arbitrary finite volume, it must have the eigenvalues $0, \pm 1, \pm 2, \ldots, \pm N, \ldots$ – even when external fields are present – (by applying the formalism of the quantization of waves). These eigenvalues now can be positive as well as negative. (In the case of the exclusion principle, the number N has an upper limit for a given size of the space region under consideration.) Besides, the charge density $\rho(x)$ and the total number of particles present are not commutable.

These requirements have now been modified with respect to the original requirements for a true relativistic single-body problem to such an extent *that no reason any longer exists for the special form* $\sum_r \psi_r^* \psi_r$ *for the charge density*. Furthermore, we shall show that the preceding conditions are fulfilled just as well in the scalar relativistic theory for spinless particles with Bose–Einstein statistics as in the Dirac hole theory. Here, Expression (3) is obviously no longer to be interpreted as a particle density but as a charge density.

The main interest of the latter theory appears to us to lie in the fact that, in this theory, the energy of the material particles is automatically always positive after the wave fields have been quantized. This is automatic in the sense that no new hypothesis equivalent to the hole idea is required and there are no concepts of passing to the limit and of subtraction, concepts which are foreign to the formalism of quantum theory.[5]

[4] If ψ_ρ^+ is the 'positive' (consisting of states of positive energy), and ψ_ρ^- is the 'negative' part of the wave function in the Dirac hole theory, the charge density as an operator has the form $\rho(x) = \sum_{\rho=1}^{1} \{\psi_\rho^{+*}\psi_\rho^+ - \psi_\rho^{-*}\psi_\rho^- + \psi_\rho^+\psi_\rho^- + \rho_\rho^{+*}\psi_\rho^{-*}\}$. Because of the appearance of the mixed terms, this expression already cannot be divided into two parts in the absence of external forces, such that each part by itself satisfies a continuity equation and forms the four-component of a four-vector.

[5] P. A. M. Dirac, *Proc. Cambr. Phil. Soc.* **30**, Pt. II, 150, 1934; R. Peierls, *Proc. Roy. Soc.* **146**, 420, 1934; W. Heisenberg, *ZS. f. Ph.* **90**, 209, 1934.

The positive value of the energy results from the fact that the Hamiltonian function of the matter–wave field, in the scalar theory discussed here, in contrast to the corresponding expression of the Dirac spinor theory, always assumes the *positive definite* form:

$$H = \int \mathrm{d}V \left\{ \left| \mathrm{i}h \frac{\partial \psi}{\partial t} - e\Phi_0 \psi \right|^2 + \sum_{k=1}^{3} \left| \mathrm{i}hc \frac{\partial \psi}{\partial x^k} + e\Phi_k \psi \right|^2 + m^2 c^4 |\psi|^2 \right\}. \tag{4}$$

This scalar relativistic theory is free of hypotheses. In the view of this, one might perhaps be surprised at first sight why 'nature has not availed itself' of this possibility of the existence of oppositely charged particles without spin and with Bose statistics, which can arise and vanish by radiation-creating or matter-creating processes.[6] But, one must consider that the question of the applicability of the theory of nuclear structure discussed here, for example to α particles, probably lies altogether outside the range of validity of present quantum theory owing to the relevant effects. Also, as will be shown in §4, the theory discussed here leads to similar infinities in connection with questions about the polarization of the vacuum, as did the original form of the hole theory.[7] It incidentally also leads to an infinite self-energy, not only of the electric particles but also to an infinite material self-energy of the light quanta.[8] Consequently, further progress in answering these questions can probably only be expected by a theoretical understanding of the numerical value of the Sommerfeld fine structure constant.

§2 Quantization of the wave field[9] in the force-free case

The Lagrange function of the scalar relativistic theory reads as follows (with $\rho, \nu, \ldots$ -1 to 4 and $x_4 = \mathrm{i}ct$):

$$L = -h^2c^2 \sum_{\nu=1}^{4} \frac{\partial \psi^*}{\partial x_\nu} \frac{\partial \psi}{\partial x_\nu} - m^2 c^4 \psi^* \psi$$

$$= h^2 \frac{\partial \psi^*}{\partial t} \frac{\partial \psi}{\mathrm{d}t} - h^2 c^2 \sum_{k=1}^{3} \frac{\partial \psi^*}{\partial x^k} \frac{\partial \psi}{\partial x^k} - m^2 c^4 \psi^* \psi. \tag{5}$$

The relativistic energy–momentum tensor becomes

$$T_{\mu\nu} = -h^2 c^2 \left(\frac{\partial \psi^*}{\partial x^\mu} \frac{\partial \psi}{\partial x^\nu} + \frac{\partial \psi^*}{\partial x^\nu} \frac{\partial \psi}{\partial x^\mu} \right) - L\delta_{\mu\nu} \tag{6}$$

[6] Compare P. A. M. Dirac, *Proc. Roy. Soc.* **133**, 60, 1931, especially p. 71.
[7] P. A. M. Dirac, Solvay Report 1933.
[8] Analogous with W. Heisenberg, *loc. cit.* – One may have doubts concerning the value of the circumstance that, in some formulations of the hole theory, the polarization effects are indeed finite but the self-energies are nevertheless infinite.
[9] As regards the expressions for the Lagrange function, energy–momentum tensor, and current vector in the scalar relativistic theory, compare e.g. W. Gordon, *Zeitschr. f. Phys.* **40**, 117, 1926.

and thus the energy (Hamiltonian function)

$$\bar{H} = \int T_{44}\,\mathrm{d}V$$
$$= \int \left\{ h^2 \frac{\partial \psi^*}{\partial t}\frac{\partial \psi}{\partial t} + h^2 c^2 \sum_{k=1}^{3} \frac{\partial \psi^*}{\partial x^k}\frac{\partial \psi}{\partial x^k} + m^2 c^4 \psi^* \psi \right\} \mathrm{d}V \tag{7}$$

and the momentum

$$G_k = \frac{\mathrm{i}}{c}\int T_{4k}\,\mathrm{d}V = -\int h^2 \left(\frac{\partial \psi^*}{\partial t}\frac{\partial \psi}{\partial x^k} + \frac{\partial \psi^*}{\partial x^k}\frac{\partial \psi}{\partial t} \right) \mathrm{d}V. \tag{8}$$

We must now regard ψ^* and ψ as q-numbers (operators acting on the Schrödinger functional), where ψ^* is the hermitian conjugate to ψ. We shall always designate the hermitian conjugate to a given q-number by means of [a superscript] $*$. We then must form the momentum π and π^* which are canonically conjugate to ψ and ψ^*, in accordance with the rule:

$$\pi = \frac{1}{h}\frac{\partial L}{\partial \left(\frac{\partial \psi}{\partial t} \right)} = h \frac{\partial \psi^*}{\partial t}, \quad \pi^* = \frac{1}{h}\frac{\partial L}{\partial \left(\frac{\partial \psi^*}{\partial t} \right)} = h \frac{\partial \psi}{\partial t}, \tag{9}$$

In the case of particles with Bose–Einstein statistics, these satisfy the canonical commutation relations (abbreviated: C.-R.):

$$\mathrm{i}[\pi(x, t), \psi(x', t)] = \delta(x - x'), \mathrm{i}[\pi^*(x, t), \psi^*(x', t)] = \delta(x - x') \tag{I}$$

where the $\delta(x - x')$ on the right side is the familiar Dirac δ-function and where, as usual, we have set

$$[A, B] = AB - BA. \tag{10}$$

The quantities ψ, ψ^* as well as π, π^* commute among themselves, as do π with ψ^* and π^* with ψ.

Application of the rule

$$\frac{\partial f}{\partial t} = \frac{\mathrm{i}}{h}[\bar{H}, f] \tag{11}$$

to ψ, ψ^* leads to an identity. Its application to π, π^*, by means of (9), leads to the wave equations

$$h^2 \frac{\partial^2 \psi}{\partial t^2} = h^2 c^2 \Delta \psi - m^2 c^4 \psi \tag{12}$$

$$h^2 \frac{\partial^2 \psi^*}{\partial t^2} = h^2 c^2 \Delta \psi^* - m^2 c^4 \psi^*. \tag{12*}$$

Further, as indeed must be the case, the rule

$$\frac{\partial f}{\partial x^k} = -\frac{\mathrm{i}}{h}[G_k, f] \tag{13}$$

is fulfilled for all quantities f. We shall see that an ambiguity in the sequence of factors occurs only in the expression for the momentum. This was chosen so that the integrand of Expression (8), which represents the momentum density, is an hermitian operator.

We now come to the expressions for the charge density ρ (measured in units of the electronic charge e), and the current density $\vec{i} = c\vec{s}$, which, together with $s_4 = \mathrm{i}\rho$, is collected together into a four-vector s. This four-vector satisfies the continuity equation

$$\sum_{\nu=1}^{4} \frac{\partial s_\nu}{\partial x_\nu} = 0 \quad \text{or} \quad \frac{\partial \rho}{\partial t} + \operatorname{div} \vec{i} = 0. \tag{14}$$

The expressions in question are given by

$$s_\nu = hci \left(\overline{\frac{\partial \psi^*}{\partial x^\nu} \psi} - \overline{\frac{\partial \psi}{\partial x^\nu} \psi^*} \right) \tag{15}$$

or

$$\rho = -hi \left(\frac{\partial \psi^*}{\partial t} \psi - \overline{\frac{\partial t}{\partial t} \psi^*} \right) = -hi \left(\frac{\partial \psi^*}{\partial t} \psi - \frac{\partial \psi}{\partial t} \psi^* \right)$$

$$= -hi \left(\psi \frac{\partial \psi^*}{\partial t} - \psi^* \frac{\partial \psi}{\partial t} \right), \tag{15a}$$

$$s_k = hci \left(\frac{\partial \psi^*}{\partial x^k} \psi - \frac{\partial \psi}{\partial x^k} \psi^* \right). \tag{15b}$$

The bars on the right side refer to the ambiguity in the sequence of factors in the expression for the density ρ, and should signify that the individual summands are to be made hermitian. It is not possible here to specify the sequence of factors solely on the basis of the requirement that the density operator should be hermitian. However, the specification made here proves to be the most suitable one, since it does not give rise to a zero-point density as will be shown later. It furthermore harmonizes with the relativistic invariance and with the continuity equation.

As can be seen, in view of (9), the density can also be written:

$$\rho = -\mathrm{i}(\pi\psi - \pi^*\psi^*) = -\mathrm{i}(\psi\pi - \psi^*\pi^*). \tag{16}$$

We shall now prove that, at a particular point x_0, this has the eigenvalues

$$\rho(x) = N \cdot \delta(x - x_0),$$

with $N = 0, \pm 1$. For this purpose, it is simplest to decompose ψ into hermitian operators

$$\psi = \frac{1}{\sqrt{2}}(u_1 + \mathrm{i}u_2), \; \psi^* = \frac{1}{\sqrt{2}}(u_1 - \mathrm{i}u_2)$$

and correspondingly

$$\pi^* = \frac{1}{\sqrt{2}}(p_1 + \mathrm{i}p_2), \; \pi = \frac{1}{\sqrt{2}}(p_1 - \mathrm{i}p_2),$$

where it follows that

$$u_1 = \frac{1}{\sqrt{2}}(\psi + \psi^*), \; u_2 = \frac{-\mathrm{i}}{\sqrt{2}}(\psi - \psi^*)$$

$$p_1 = h\frac{\partial u_1}{\partial t} = \frac{1}{\sqrt{2}}(\pi^* + \pi), \; p_2 = h\frac{\partial u_2}{\partial t} = \frac{-\mathrm{i}}{\sqrt{2}}(\pi^* - \pi).$$

Then one has

$$\mathrm{i}[p_1(x), u_1(x')] = \delta(x - x'), \; \mathrm{i}[p_2(x), u_2(x)] = \delta(x - x'),$$

while every quantity with index 1 commutes with every quantity with index 2. Since the energy, momentum, and wave equations decompose additively into expressions which depend only on p_1, u_1 or respectively p_2, u_2, one then has

$$\rho = p_1 u_2 - p_2 u_1. \tag{16a}$$

Due to the analogy with the expression for a component of the angular momentum, it is immediately obvious from this that $\rho(x)$ has the eigenvalue $N \cdot \delta(x - x')$ with $N = 0, \pm 1, \ldots$. (The factor $\delta(x - x')$ can here be justified, for example, by a limiting process from a discrete subdivision of space into a continuous one.) Since the values of the density at various positions commute with one another, it does in fact follow that the charge situated within an arbitrary finite region,

$$e_v = \int_v \rho \, dV$$

(measured in units of e), has the eigenvalues $0, \pm 1, \ldots, \pm N$.

We also note that, in the present theory, all relations including the C.-R. remain correct, if all operators are exchanged with their hermitian conjugates (that is ψ with ψ^*, π with π^*). Since a four-current here changes its sign, this shows the symmetry of the theory with respect to positive and negative charges.

Incidentally, it should also be noted here that a decomposition of the density ρ into commutable parts with only positive and only negative eigenvalues is indeed possible in infinitely many ways, but that none of these parts by themselves satisfy a continuity equation and also are not relativistically invariant.[10]

We shall now investigate the behavior in momentum space. This is important for applications and it is also intrinsically of physical interest. To obtain sums instead of integrals in momentum space, we use the familiar formal method of imposing on the wave fields the condition of having as a periodicity region a cube of length L, that is with volume $L^3 = V$, so that the components of the wave vector $\vec{k}$ must be integer multiples of $2\pi/L$. We further use

$$u_k = \frac{1}{V^{1/2}} \exp \mathrm{i}(\vec{k} \cdot \vec{x}) \tag{17}$$

as a complete system of orthogonally normalized c-number eigenfunctions. For these eigenfunctions there thus holds

$$\int_V u_k^*(x) u_l(x) \, dV = \delta_{kl}. \tag{18}$$

For the sake of simplicity, we shall write, both here and below, only one index as the index k instead of the three indices corresponding to the three components of $\vec{k}$, and similarly for the sums over k.

We now decompose the functions ψ, π, ψ^*, π^* in terms of the u_k according to

$$\psi = \frac{1}{V^{1/2}} \sum_k q_k \exp \mathrm{i}(\vec{k} \cdot \vec{x}), \; \psi^* = \frac{1}{V^{1/2}} \sum_k q_k^* \exp -\mathrm{i}(\vec{k} \cdot \vec{x}) \tag{19a}$$

[10] Such decompositions are obtained e.g. by introducing the arbitrary constant a with the dimension corresponding to the square root of the energy (e.g. $a = (mc^2)^{1/2}$, according to the formulation $\pi = (a/\sqrt{2})(\varphi_1 + \varphi_2^*)$, $\psi = (-\mathrm{i}/\sqrt{2}a)(\varphi_1^* - \varphi_2)$, $\pi^* = (a/\sqrt{2})(\varphi_1^* + \varphi_2)$, $\psi^* = (-\mathrm{i}/\sqrt{2}a)(\varphi_2^* - \varphi_1)$. One then has $[\varphi_1(x), \varphi_1^*(x')] = \delta(x - x')$, $[\varphi_2(x), \varphi_2^*(x')] = \delta(x - x')$, where quantities with index 1 and quantities of index 2 commute. One also has $\rho = \varphi_2^* \varphi_2 - \varphi_1^* \varphi_1$. This furnishes another proof for the eigenvalues of ρ.

$$\pi^* = \frac{1}{V^{1/2}} \sum_k p_k^* \exp \mathrm{i}(\vec{k} \cdot \vec{x}), \; \pi = \frac{1}{V^{1/2}} \sum_k p_k \exp -\mathrm{i}(\vec{k} \cdot \vec{x}) \tag{19b}$$

with the inverse formulas

$$q_k = \frac{1}{V^{1/2}} \int_V \psi \exp -\mathrm{i}(\vec{k} \cdot \vec{x})\, \mathrm{d}V, \; q_k^* = \frac{1}{V^{1/2}} \int_V \psi^* \exp \mathrm{i}(\vec{k} \cdot \vec{x})\, \mathrm{d}V \tag{19c}$$

$$p_k^* = \frac{1}{V^{1/2}} \int_V \pi^* \exp -\mathrm{i}(\vec{k} \cdot \vec{x})\, \mathrm{d}V, \; p_k = \frac{1}{V^{1/2}} \int_V \pi \exp \mathrm{i}(\vec{k} \cdot \vec{x})\, \mathrm{d}V. \tag{19d}$$

Then the q-numbers p_k, q_k, p_k^*, q^*, (one should note that p_k, q_k are not hermitian operators and p_k^*, q_k^* are the hermitian conjugates of $p_k q_k$) satisfy the C.-R.

$$\mathrm{i}[p_k, q_l] = \delta_{kl}, \; \mathrm{i}[p_k^*, q_l^*] = \delta_{kl}, \tag{II}$$

in accordance with (I) while the q_k and q_l^* commute with one another, as do the p_k and p_l^*, as well as the p_k with the q_l^*, and the p_k^* with the q_l. Furthermore, according to (9) we have

$$p_k = h\dot{q}_k^*, \; p_k^* = h\dot{q}_k. \tag{20}$$

According to (7) and (8), one obtains for the Hamiltonian function and for the momentum

$$\bar{H} = \sum_k (p_k^* p_k + E_k^2 q_k^* q_k) \tag{21}$$

$$\vec{G} = -\mathrm{i}h \sum_k \vec{k}(p_k q_k - q_k^* p_k^*). \tag{22}$$

As an abbreviation, we here set

$$E_k^2 = c^2(h^2 k^2 + m^2 c^2). \tag{23}$$

Below, we shall always take

$$E_k = + c(h^2 k^2 + m^2 c^2)^{1/2} \tag{23a}$$

to mean the positive root.

The validity of rule (II) for p_k, q_k, p_k^*, q_k^* is easily confirmed; in particular one has

$$\dot{p}_k = \frac{\mathrm{i}}{h}[\bar{H}, p_k] = -\frac{\mathrm{i}}{h} E_k^2 q_k^*, \tag{24a}$$

$$\dot{p}_k^* = \frac{\mathrm{i}}{h}[\bar{H}, p_k^*] = -\frac{1}{h} E_k^2 q_k. \tag{24b}$$

Because of (16) and (15b), we can also write the expressions for the total charge,

$$\bar{e} = \int_V \rho \, \mathrm{d}V,$$

and the total current,

$$\frac{1}{c}\vec{J} = \int_V \vec{s}\, \mathrm{d}V,$$

in their decomposition in terms of components of the various momentum eigenfunctions. We obtain

$$\bar{e} = -\mathrm{i} \sum_k (p_k q_k - p_k^* q_k^*), \tag{25}$$

$$\frac{1}{c}\vec{J} = 2hc\sum_k \vec{k} q_k^* q_k. \tag{26}$$

We shall see that the latter is not constant in time.

We shall now show that the components of the individual proper vibration k belonging to the total charge, the energy, and the momentum, can be decomposed simultaneously into two parts which allow a simple physical interpretation. For this purpose, we introduce the following variables a_k, a_k^*, b_k, b_k^*:

$$p_k = \left(\frac{E_k}{2}\right)^{1/2}(a_k^* + b_k),\ q_k = \frac{-\mathrm{i}}{(2E_k)^{1/2}}(-a_k + b_k^*), \tag{27}$$

$$p_k^* = \left(\frac{E_k}{2}\right)^{1/2}(a_k + b_k^*),\ q_k^* = \frac{-\mathrm{i}}{(2E_k)^{1/2}}(a_k^* - b_k), \tag{27*}$$

with the inversion formulas

$$a_k = \frac{1}{2^{1/2}}\left(\frac{1}{E_k^{1/2}}p_k^* - E_k^{1/2}q_k\right),\ a_k^* = \frac{1}{2^{1/2}}\left(\frac{1}{E_k^{1/2}}p_k + \mathrm{i}E_k^{1/2}q_k^*\right), \tag{28a}$$

$$b_k = \frac{1}{2^{1/2}}\left(\frac{1}{E_k^{1/2}}p_k - \mathrm{i}E_k^{1/2}q_k^*\right),\ b_k^* = \frac{1}{2^{1/2}}\left(\frac{1}{E_k^{1/2}}p_k^* + \mathrm{i}E_k^{1/2}q_k\right). \tag{28b}$$

The commutation relations

$$[a_k, a_l^*] = \delta_{kl},\ [b_k, b_l^*] = \delta_{kl}, \tag{III}$$

follow for the new variables, while the a_k or a_k^* commute with each other, as do the b_k and b_k^*, the a_k with the b_l^*, and the a_k^* with the b_l.

One further obtains

$$\bar{H} = \sum_k E_k \tfrac{1}{2}(a_k a_k^* + a_k^* a_k + b_k^* b_k + b_k b_k^*)$$

$$= \sum_k E_k(a_k^* a_k + b_k^* b_k + 1), \tag{29}$$

$$\vec{G} = h\sum_k \vec{k}\tfrac{1}{2}(a_k^* a_k + a_k a_k^* - b_k^* b_k - b_k b_k^*)$$

$$= h\sum_k \vec{k}(a_k^* a_k - b_k^* b_k). \tag{30}$$

Further, for the total charge

$$\bar{e} = \sum_k \tfrac{1}{2}(a_k^* a_k + a_k a_k^* - b_k^* b_k - b_k b_k^*)$$

$$= \sum_k (a_k^* a_k - b_k^* b_k). \tag{31}$$

Finally, according to (26), one obtains for the total current

$$\frac{1}{c}\vec{J} = hc\sum_k \frac{\vec{k}}{E_k}(a_k^* a_k + b_k b_k^* - a_k^* b_k^* - a_k b_k)$$

$$= hc\sum_k \frac{\vec{k}}{E_k}(a_k^* a_k + b_k^* b_k - a_k^* b_k^* - a_k b_k + 1). \tag{32}$$

The C.-R.s for the a, b, a^*, b^* have as their consequence that the operators

$$N_k^+ = a_k^* a_k,\ N_k^- = b_k^* b_k \tag{33}$$

commute and both of them have the non-negative integer eigenvalues 0, 1, 2, ... The expressions for the charge, energy, and momentum justify the following interpretation in the force-free case which is being considered here:

N_k^+ *signifies the number of particles with charge number +1 and momentum* $h\vec{k}$, *and* N_k^- *signifies the number of particles with charge number −1 and momentum* $-h\vec{k}$.[11]

It should also be pointed out that the term with +1 in the energy expression denotes a zero-point energy (vacuum energy) of the matter waves. However, quite analogous to the zero-point energy of the electromagnetic radiation, this can be deleted in all applications and regardless of the relativistic invariance of the theory. The like holds for the term with +1 in the expression for the current. Even apart from this term, the energy is automatically always positive, and this has decisive physical significance.

The terms with $a_k b_k$ and $a_k^* b_k^*$ in the expression for the current are important. These terms prevent constancy in time even in the force-free case. As can be seen from the equations of motion

$$\dot{a}_k = -\mathrm{i}\frac{E_k}{h}a_k,\quad \dot{b}_k = -\mathrm{i}\frac{E_k}{h}b_k \tag{34}$$

and their integrals

$$a_k = a_k(0)\exp\left(-\mathrm{i}\frac{E_k}{h}t\right),\ b_k = b_k(0)\exp\left(-\mathrm{i}\frac{E_k}{h}t\right), \tag{35}$$

$$a_k^* = a_k^*(0)\exp\left(+\mathrm{i}\frac{E_k}{h}t\right),\ b_k^* = b_k^*(0)\exp\left(+\mathrm{i}\frac{E_k}{h}t\right), \tag{35*}$$

which are obtained from (III) and (29) in accord with (II), these terms have a close analogy to the Schrödinger trembling motion. As will be shown in the following paragraph, these terms actually cause processes of pair creation and pair annihilation in the presence of suitable external fields.

As already mentioned in the introduction, it is not possible to carry through the scalar relativistic wave theory consistently for particles obeying the exclusion principle. As a more detailed investigation of the Hamiltonian function with the variables a and b shows, the reason for this is as follows: When Fermi statistics are valid, the relativistic invariance of the four-current cannot be achieved. This is also connected with the fact that $\psi(x) = 0$ and $\psi^*(x) = 0$ would follow from the equations

$$\psi(x)\psi^*(x') + \psi^*(x')\psi(x) = 0,\quad \psi(x)\psi(x) + \psi(x')\psi(x) = 0.$$

§3 The case where external forces are present

For a particle with charge e, one goes from the force-free case to the case of the presence of an external electromagnetic field with the four-potential $\Phi_\mu(\Phi_4 = \mathrm{i}\Phi_0)$, by

[11] It should again be noted that a corresponding definition for a spatial density $\rho^+(x)$ and $\rho^-(x)$ of the particle types is not possible in a physically meaningful fashion. For example, if one forms $a(x) = V^{-1/2}\sum_k a_k \exp\mathrm{i}(\vec{k}\cdot\vec{x})$, $b(x) = V^{-1/2}\sum_k b_k \exp -\mathrm{i}(\vec{k}\cdot\vec{x})$ from a_k and b_k, it appears that the expression $a^*(x)a(x) - b^*(x)b(x)$ does *not* agree with the charge density.

replacing the operator p_μ by

$$p_\mu \to p_\mu - \frac{e}{c}\Phi_\mu, \tag{36}$$

which corresponds[12] to the substitutions

$$\frac{\partial\psi}{\partial x^\mu} \to \frac{\partial\psi}{\partial x^\mu} - \frac{ie}{hc}\Phi_\mu\psi, \quad \frac{\partial\psi^*}{\partial x^\mu} \to \frac{\partial\psi^*}{\partial x^\mu} + \frac{ie}{hc}\Phi_\mu\psi^* \tag{36a}$$

The Lagrange function of the matter field then becomes

$$\begin{aligned} L &= -h^2c^2\sum_{\nu=1}^{4}\left(\frac{\partial\psi^*}{\partial x^\nu} + \frac{ie}{hc}\Phi_\nu\psi^*\right)\left(\frac{\partial\psi}{\partial x^\nu} - \frac{ie}{hc}\Phi_\nu\psi\right) - m^2c^4\psi^*\psi \\ &= h^2\left(\frac{\partial\psi^*}{\partial t} - \frac{ie}{h}\Phi_0\psi^*\right)\left(\frac{\partial\psi}{\partial t} + \frac{ie}{h}\Phi_0\psi\right) \\ &\quad -h^2c^2\sum_{k=1}^{3}\left(\frac{\partial\psi^*}{\partial x^k} + \frac{ie}{hc}\Phi_k\psi^*\right)\left(\frac{\partial\psi}{\partial x^k} - \frac{ie}{hc}\Phi_k\psi\right) - m^2c^4\psi^*\psi. \end{aligned} \tag{37}$$

The Hamiltonian function of the matter field becomes

$$\begin{aligned} \bar{H}^{\mathrm{m}} = \int\Bigg\{&\left(h\frac{\partial\psi^*}{\partial t} - \mathrm{i}e\Phi_0\psi^*\right)\left(h\frac{\partial\psi}{\partial t} + \mathrm{i}e\Phi_0\psi\right) \\ &+ \sum_{k=1}^{3}\left(hc\frac{\partial\psi^*}{\partial x^k} + \mathrm{i}e\Phi_k\psi^*\right)\left(hc\frac{\partial\psi}{\partial x^k} - \mathrm{i}e\Phi_k\psi\right) + m^2c^4\psi^*\psi\Bigg\}\,\mathrm{d}V. \end{aligned} \tag{37a}$$

If one adds to this the expressions for the electromagnetic Lagrange function

$$L^{\mathrm{elm}} = \frac{1}{8\pi}(E^2 - H^2), \tag{38}$$

and respectively for the electromagnetic energy

$$\bar{H}^{\mathrm{elm}} = \frac{1}{8\pi}\int(E^2 + H^2)\,\mathrm{d}V, \tag{38a}$$

one obtains the energy integral

$$\bar{H}^{\mathrm{m}} + \bar{H}^{\mathrm{elm}} = \mathrm{const.}$$

Furthermore, by variation of the action integral

$$\int(L^{\mathrm{m}} + L^{\mathrm{elm}})\,\mathrm{d}V\,\mathrm{d}t$$

in terms of the field variables ψ, ψ^*, Φ_μ, one obtains on the one hand the wave equations

$$\begin{aligned} &\left(h\frac{\partial}{\partial t} - \mathrm{i}e\Phi_0\right)\left(h\frac{\partial}{\partial t} - \mathrm{i}e\Phi_0\right)\psi \\ &= \sum_{k=1}^{3}\left(hc\frac{\partial}{\partial x^k} + \mathrm{i}e\Phi_k\right)\left(hc\frac{\partial}{\partial x^k} + \mathrm{i}e\Phi_k\right)\psi + m^2c^4\psi, \end{aligned} \tag{39}$$

[12] In the Dirac theory, the substitution $p_\mu \to p_\mu + (e/c)\Phi_\mu$ is introduced, since the electron charge is there designated by $(-e)$. Our designations agree with W. Gordon, *loc. cit.*

$$\left(h\frac{\partial}{\partial t}+\mathrm{i}e\Phi_0\right)\left(h\frac{\partial}{\partial t}+\mathrm{i}e\Phi_0\right)\psi^*$$
$$=\sum_{k=1}^{3}\left(hc\frac{\partial}{\partial x^k}-\mathrm{i}e\Phi_k\right)\left(hc\frac{\partial}{\partial x^k}-\mathrm{i}e\Phi_k\right)\psi^*+m^2c^4\psi^*, \tag{39*}$$

and on the other hand Maxwell's equations

$$\mathrm{rot}\,\vec{H}-\frac{1}{c}\dot{\vec{E}}=4\pi e\frac{1}{c}\vec{i}=4\pi e\vec{s}, \tag{40}$$
$$\mathrm{div.}\,\vec{E}=4\pi e\rho, \tag{41}$$

with the following expressions for ρ and $\vec{s}$

$$\rho=\mathrm{i}\left[\left(h\frac{\partial\psi}{\partial t}+\mathrm{i}e\Phi_0\psi\right)\psi^*-\left(h\frac{\partial\psi^*}{\partial t}-\mathrm{i}e\Phi_0\psi^*\right)\psi\right]$$
$$=h\mathrm{i}\left(\frac{\partial\psi}{\partial t}\psi^*-\frac{\partial\psi^*}{\partial t}\psi\right)-2e\Phi_0\psi^*\psi, \tag{42}$$
$$s_k=\mathrm{i}\left[\left(hc\frac{\partial\psi^*}{\partial x^k}+\mathrm{i}e\Phi_k\psi^*\right)\psi-\left(hc\frac{\partial\psi}{\partial x^k}-\mathrm{i}e\Phi_k\psi\right)\psi^*\right]$$
$$=\mathrm{i}hc\left(\frac{\partial\psi^*}{\partial x^k}\psi-\frac{\partial\psi}{\partial x^k}\psi^*\right)-2e\Phi_k\psi^*\psi. \tag{43}$$

These differ from the corresponding expressions in the force-free case (15a) and (15b) by characteristic additions. The validity of the continuity equation (14) for the new expressions for ρ and s_k follows both from the wave equations (39) and from the Maxwell equations. A direct consequence of (36) is the invariance of the Lagrange and Hamiltonian functions and also of the expressions for the current and charge densities with respect to the gauge transformations

$$\Phi'_\mu=\Phi_\mu+\frac{\partial\lambda}{\partial x^\mu},\quad \psi'=\psi\exp\left(\mathrm{i}\frac{e}{hc}\lambda\right). \tag{36b}$$

It is also important that the wave equations, the Maxwell equations, and the Hamiltonian function remain intact if ψ is exchanged with ψ^* and at the same time e is replaced by $-e$. The consequence of this is that the theory is symmetric with respect to positive and negative charge. All these statements remain intact if one regards ψ, ψ^*, Φ_μ as q-numbers.

It is important that the meaning of π, π^* is changed as compared to (9), in accordance with

$$\pi=\frac{1}{h}\frac{\partial L}{\partial\left(\frac{\partial\psi}{\partial t}\right)}=h\frac{\partial\psi^*}{\partial t}-\mathrm{i}e\Phi_0\psi^* \tag{44}$$
$$\pi^*=\frac{1}{h}\frac{\partial L}{\partial\left(\frac{\partial\psi^*}{\partial t}\right)}=h\frac{\partial\psi}{\partial t}+\mathrm{i}e\Phi_0\psi. \tag{44*}$$

These new π now satisfy the C.-R. (I)

$$\mathrm{i}[\pi(x,t),\psi(x',t)]=\delta(x-x'),\ \mathrm{i}[\pi^*(x,t),\psi^*(x',t)]=\delta(x-x') \tag{I}$$

and can be commuted with the electromagnetic field variables. According to (42), the charge density then becomes

$$\rho = \mathrm{i}(\pi^* \psi^* - \pi\psi) = \mathrm{i}(\psi^* \pi^* - \psi\pi), \tag{16}$$

again in formal agreement with (16). *Consequently, the eigenvalues of the charge density remain the same as in the force-free case, even when external potentials are present.*

The material part (37a) of the Hamiltonian function is now written

$$\bar{H}^{\mathrm{m}} = \bar{H}_0 + \bar{H}_1,$$

with

$$\bar{H}_0 = \int \left\{ \pi\pi^* + h^2 c^2 \sum_{k=1}^{3} \frac{\partial \psi^*}{\partial x^k} \frac{\partial \psi}{\partial x^k} + m^2 c^4 \psi^* \psi \right\} \mathrm{d}V, \tag{45_0}$$

$$\bar{H}_1 = \int \left\{ ehc \sum_{k=1}^{3} \Phi_k \left(\psi^* \frac{\partial \psi}{\partial x^k} - \frac{\partial \psi^*}{\partial x^k} \psi \right) + e^2 \sum_{k=1}^{3} \Phi_k^2 \psi^* \psi \right\} \mathrm{d}V. \tag{45_1}$$

With the familiar C.-R.s for the field strengths and electromagnetic potentials, quantum electrodynamics can also be formulated in the familiar fashion, where the Maxwell equations (40) can be obtained by applying rule (II) according to

$$\frac{\partial \vec{E}}{\partial t} = \frac{\mathrm{i}}{h} [\bar{H}, \vec{E}]$$

We do not wish to discuss this in more detail, but just recall the familiar complications which result from the fact that Equation (41) is commutable only with the gauge-invariant quantities $\psi\pi$, $\psi^*\pi^*$, $\vec{E}$, $\vec{H}$, but not with others such as π, π^*, ψ, ψ^*, Φ_μ. If rule (II)

$$\frac{\partial f}{\partial t} = \frac{\mathrm{i}}{h} [H, f]$$

is also to be valid for these quantities, one must formally add an expression of the form

$$\int \Phi_0 (4\pi e \rho - \mathrm{div.}\, \vec{E})\, \mathrm{d}V$$

to the sum of (38a), (45_0), and (45_1).[13]

For the part H' of the Hamiltonian function, which does not agree with (37a),

$$H' = \bar{H}_0 + \bar{H}_1 + \bar{H}_2 \tag{47}$$

$$\bar{H}_2 = e \int \Phi_0 \rho \, \mathrm{d}V = \mathrm{i}e \int \Phi_0 (\pi^* \psi^* - \pi\psi)\, \mathrm{d}V \tag{45_2}$$

we thus have

$$\dot{\pi} = \frac{\mathrm{i}}{h} [H', \pi], \quad \dot{\psi} = \frac{\mathrm{i}}{h} [H', \psi], \tag{48}$$

and the corresponding relations for the canonically conjugate π^*, ψ^*.

This (not gauge-invariant) Hamiltonian function H' consequently also has the property of being constant in time if the potentials Φ_0 and Φ_k do indeed depend on the space coordinates but not on time. For many purposes, it is sufficient to consider the four-potentials as given c-number functions. In this case, one must calculate with the

[13] Compare also W. Heisenberg and W. Pauli, *Zeitschr. f. Phys.* **59**, 168, 1930; especially p. 179, Equation (38).

Hamiltonian function given by H'. Incidentally, this is also obtained by canonically transforming L^{m} according to the formula

$$H' = \int \pi \dot{\psi} \,\mathrm{d}V + \int \pi^* \dot{\psi}^* \,\mathrm{d}V - L^{\mathrm{m}}.$$

We shall now also write down the additional parts H_1 and H_2 of the Hamiltonian function in momentum space, as has already been done in the previous paragraph with the original part H_0 [compare Equation (21)]. The matrix elements of a function $f(x)$ (e.g. the potentials) are defined with reference to the orthogonal system of functions (17) in their usual fashion:

$$f_{kl} = \frac{1}{V} \int f(\vec{x}) \exp(-\mathrm{i}[(\vec{k} - \vec{l})\vec{x}]) \,\mathrm{d}V. \tag{49}$$

Then, from (19a), (19b) we obtain directly

$$\overline{H}_2 = \mathrm{i}e \sum_l \sum_k \Phi^o_{kl} (p_l^* q_k^* - p_k q_l) \tag{50}$$

$$\overline{H}_1 = -\sum_l \sum_k [hce(\vec{\Phi}_{kl}, \vec{k} + \vec{l}) - e^2 (\vec{\Phi})^2_{kl}] q_k^* q_l \tag{51}$$

or, by introducing the variables a_k, b_k according to (27), (28), in which H_0 is given by (29):

$$\overline{H}_2 = \tfrac{1}{2} e \sum_l \sum_k \Phi^o_{kl} \left[\frac{E_k + E_l}{(E_k E_l)^{1/2}} (a_k^* a_l - b_l^* b_k) + \frac{E_k - E_l}{(E_k E_l)^{1/2}} (a_l b_k - a_k^* b_l^*) \right] \tag{52}$$

$$\overline{H} = \tfrac{1}{2} \sum_l \sum_k \frac{1}{(E_k E_l)^{1/2}} [hce(\vec{\Phi}_{kl}, \vec{k} + \vec{l}) - e^2 (\vec{\Phi})^2_{kl}] (a_k^* a_l + b_k b_l^* - a_k^* b_l^* - b_k a_l). \tag{53}$$

§4 Pair creation by light quanta and the polarization of the vacuum

From the commutation relations (III) for the a_k^*, a_k and respectively b_k^*, b_k the properties of these operators follow in familiar fashion when they are applied to a Schrödinger function $c(\ldots N_k^+ \ldots; \ldots N_k^- \ldots)$ which depends on the occupation numbers N_k^+, N_k^-. One has

$$a_k^* c(\ldots N_k^+ \ldots; \ldots N_k^- \ldots) = (N_k^+ + 1)^{1/2} c(\ldots N_k^+ + 1 \ldots; \ldots N_k^- \ldots)$$

$$a_k c(\ldots N_k^+ \ldots; \ldots N_k^- \ldots) = (N_k^+)^{1/2} c(\ldots N_k^+ - 1 \ldots; \ldots N_k^- \ldots)$$

$$b_k^* c(\ldots N_k^+ \ldots; \ldots N_k^- \ldots) = (N_k^- + 1)^{1/2} c(\ldots N_k^+ \ldots; \ldots N_k^- + 1 \ldots)$$

$$b_k c(\ldots N_k^+ \ldots; \ldots N_k^- \ldots) = (N_k^-)^{1/2} c(\ldots N_k^+ \ldots; \ldots N_k^- - 1 \ldots). \tag{54}$$

One can then easily see that the additions $\overline{H}_1$ and $\overline{H}_2$, which appear in the Hamiltonian function when external fields are present, contain terms due to the factors $a_k^* b_l^*$ and $b_k a_l$, which cause pair creation and pair annihilation. These terms do indeed lead to matrix elements between the states which differ precisely by one positive and one negative particle, while the factors $a_k^* a_l$ and respectively $b_k b_l^*$ yield only transitions of one positive or respectively one negative charge from one state to another.

Below we shall now calculate the probability of pair creation through a light quantum of energy $h\nu > 2mc^2$, on the basis of expressions (52) and (53). We shall compare this probability with the corresponding expressions of the hole theory, as calculated by Bethe and Heitler.[14]

Due to the law of conservation of energy and momentum, this probability vanishes in field-free space. We therefore assume that an electric field exists in space, a field which can be represented by a time-independent scalar potential Φ_0 (for example the Coulomb field of a nucleus), which can take up the momentum excess.

We consider the influence of the field only in first approximation, just like Bethe and Heitler, by starting from field-free space and by considering Φ_0 as well as the potential of the light wave as a perturbation.

If we now ask about the probability W that, per unit time, a positive and a negative particle with momenta $h\vec{k}$ and $-h\vec{l}$ respectively, and with energy E_k and E_l, respectively, are created through the absorption of a light quantum $h\nu = E_k + E_l$ in empty space. Only in second approximation do we then obtain a non-vanishing result:

$$W = \frac{1}{h^2}\left|\sum_c \frac{\bar{H}_1(AC)\bar{H}_2(CB)}{E_B - E_C} + \frac{\bar{H}_2(AC)\bar{H}_1(CB)}{E_A - E_C}\right|^2. \tag{55}$$

Let A be the vacuum state (all $N = 0$), let B be the final state ($N_k^+ = 1$, $N_l^- = 1$, all other $N = 0$), and let C denote any intermediate state. $\bar{H}_1(AC)$ . . . etc. signifies the matrix element of $\bar{H}_1$ between the states A and C, where one must insert in $\bar{H}_2$ the scalar potential Φ_0 and in $\bar{H}_1$ the vector potential $\vec{\Phi}$ of the light wave with frequency ν. Due to conservation of momentum, which appears in the calculation of $\bar{H}_1$, only the following four intermediate states are relevant:

$$\left.\begin{array}{l} C_1 \rightarrow N_k^+ = 1,\ N_{k-n}^- = 1 \\ C_2 \rightarrow N_{l+n}^+ = 1,\ N_l^- = 1 \\ C_3 \rightarrow N_k^+ = 1,\ N_{l+n}^- = 1 \\ C_4 \rightarrow N_{k-n}^+ = 1,\ N_l^- = 1 \end{array}\right\} \quad \text{all other } N = 0.$$

The corresponding matrix elements can be calculated from (52) and (53) and lead to an expression which we write down as the differential cross section $\mathrm{d}Q$, that an unpolarized light quantum of frequency ν will create a positive particle with energy between E_+ and $E_+ + \mathrm{d}E$ and a negative particle with energy between E_- and $E_- - \mathrm{d}E$ ($E_+ + E_- = h\nu$), and that their momenta $\vec{p}_+$ and $\vec{p}_-$ will make the angles θ_+ and θ_- with the direction of the light quantum:

$$\mathrm{d}Q = \frac{1}{8\pi^3}\frac{\mathrm{e}^2}{hc}\frac{1}{h^3\nu^3}\sin\theta_+\sin\theta_-\,\mathrm{d}\theta_+\mathrm{d}\theta_-\,\mathrm{d}\varphi\,\frac{p_+p_-}{h^4}\,\mathrm{d}E|\Phi_0(\vec{q})|^2$$
$$\left\{\frac{E_-^2p_+^2\sin^2\theta_+}{(E_+ - cp_+\cos\theta_+)^2} + \frac{E_+^2p_-^2\sin^2\theta_-}{(E_- - cp_-\cos\theta_-)^2} + \frac{2E_+E_-p_+p_-\sin\theta_+\sin\theta_-\cos\varphi}{(E_+ - cp_+\cos\theta_+)(E_- - cp_-\cos\theta_-}\right\}. \tag{56}$$

[14] *Proc. Roy. Soc.* **146**, 83, 1934.

θ is the angle between the planes which are formed from the direction of the light quantum and the directions p_+ and p_- respectively.

$\Phi_0(\vec{q})$ here designates the matrix element

$$\Phi_0(\vec{q}) = \int \Phi_0(x) \exp(\mathrm{i}(\vec{q} \cdot \vec{x}))\, \mathrm{d}V$$

(in contrast to (49), this no longer contains the total volume V), where $h\vec{q}$ is the momentum excess delivered to the electric field:

$$h\vec{q} = (\vec{p}_+ - \vec{p}_- - h\vec{n}),$$

and $\vec{n}$ is the wave vector of the light quantum.

In the Coulomb field $\Phi_0 = Ze/r$, one must set:

$$\Phi_0(\vec{q}) = 4\pi Ze \frac{1}{q^4}.$$

The corresponding expression in hole theory, according to Bethe and Heitler, reads as follows:

$$\begin{aligned} \mathrm{d}Q = {} & \frac{1}{8\pi^3} \frac{\mathrm{e}^2}{hc} \frac{1}{h^3 v^3} \sin\theta_+ \sin\theta_- \,\mathrm{d}\theta_+ \,\mathrm{d}\theta_- \,\mathrm{d}\varphi \frac{p_+ p_-}{h^4} \,\mathrm{d}E |\Phi_0(\vec{q})|^2 \\ & \cdot \left\{ \frac{p_+^2 \sin_+^2 \theta_+ (E_-^2 - h^2 c^2 q^2/4)}{(E_+ - cp_+ \cos\theta_+)^2} + \frac{p_-^2 \sin_-^2 \theta_- (E_+^2 - h^2 c^2 q^2/4)}{(E_- - cp_- \cos\theta_-)^2} \right. \\ & + \frac{2p_- p_+ \sin\theta_- \sin\theta_+ \cos\varphi (E_- E_+ + c^2 h^2 q^2/4)}{(E_+ - cp_+ \cos\theta_+)(E_- - cp_- \cos\theta_-)} \\ & \left. - \frac{\frac{1}{2} h^2 v^2 [p_+^2 \sin^2\theta_+ + \mathrm{p}_-^2 \sin^2\theta_- + 2p_+ p_- \sin\theta_+ \sin\theta_- \cos\theta]}{(E_+ - cp_+ \cos\theta_+)(E_- - cp_- \cos\theta_-)} \right\}. \end{aligned}$$

This differs from expression (56), which was obtained from the scalar wave equation, only by the third term in curly brackets and in the terms with q^2 that appear there. However, at high energies the latter are negligible, since $hc|\vec{q}| \ll h\nu$ for $h\nu \gg mc^2$.

If one inserts the Coulomb potential for Φ_0, the integration over the angles can easily be performed for the limiting case $h\nu \gg mc^2$.[15] One obtains

$$\mathrm{d}Q = \frac{Z^2 e^2}{hc} \left(\frac{e^2}{mc^2} \right)^2 \frac{32}{3} \frac{E_+ E_-}{h^3 v^3} \left(\lg \frac{2E_+ E_-}{h\nu mc^2} - \frac{1}{2} \right)$$

and for the total cross section:

$$Q = \frac{Z^2 e^2}{hc} \left(\frac{e^2}{mc^2} \right)^2 \left(\frac{16}{9} \lg \frac{2h\nu}{mc^2} - \frac{104}{27} \right).$$

According to Bethe and Heitler, the corresponding formulas of hole theory read as follows:

$$\mathrm{d}Q = \frac{Z^2 e^2}{hc} \left(\frac{e^2}{mc^2} \right)^2 4 \frac{E_+^2 + E_-^2 + \frac{2}{3} E_+ E_-}{h^3 v^3} \left(\lg \frac{2E_+ E_-}{h\nu mc^2} - \frac{1}{2} \right),$$

[15] We owe a debt of thanks to Mr Bethe for entrusting to us the manuscript of his paper which will appear in *Proc. Cambr. Phil. Soc.*, in which similar integrations are performed. [See *Proc. Camb. Phil. Soc.* 30: 524 (1934).]

and respectively

$$Q = \frac{Z^2e^2}{hc}\left(\frac{e^2}{mc^2}\right)^2\left(\frac{28}{9}\lg\frac{2h\nu}{mc^2} - \frac{218}{27}\right).$$

The cross section for pair creation, in the theory treated here, is thus smaller by about a factor 4/7 in the limit $h\nu \gg mc^2$.

Finally, the polarization of the vacuum through an electrostatic field will be calculated. For this purpose, we calculate the additional charge density $\rho(x)$ which results from the field Φ_0 of an 'external' charge density $\rho_0(x)$ which is present in space; $\rho(x)$ is then the charge density induced by a potential Φ_0, in a space which is *empty* as regards the positive and negative particles described by the wave equation.

It is advantageous to Fourier decompose the density. From (16) and (19), one then obtains for the Fourier coefficients:

$$\rho(\vec{\zeta}) = \frac{1}{V}\int \rho(\vec{x})\exp(-\mathrm{i}(\vec{\zeta}\cdot\vec{x}))\,\mathrm{d}V = \sum_k (p_l^* q_k^* - p_k q_l),$$

where $\vec{l} = \vec{k} + \vec{\zeta}$. Further, quoting (27), and by introducing the operators a and b, one obtains

$$\rho(\vec{\zeta}) = \frac{1}{2}\sum_k \left\{\frac{E_k - E_l}{(E_kE_l)^{1/2}}(a_k^*\alpha_l - b_l^*b_k) + \frac{E_k + E_l}{(E_kE_l)^{1/2}}(a_k^*b_l^* - a_lb_k)\right\}. \quad (57)$$

Now we must apply this operator to the Schrödinger functional of empty space $c(\ldots 0 \ldots; \ldots 0 \ldots)$, which is perturbed by the external field Φ_0. A first-order perturbation equation yields, from (52):

$$\begin{aligned} c(\ldots 0 \ldots; \ldots 0 \ldots) &= c_0(\ldots 0 \ldots; \ldots 0 \ldots) \\ &- \frac{1}{2}\sum_{kl}\Phi^0_{kl}\frac{E_k - E_l}{(E_kE_l)^{1/2}(E_k + E_l)}c_0(\ldots 1_k \ldots; \ldots 1_l \ldots), \end{aligned} \quad (58)$$

where $c_0(\ldots 0 \ldots; \ldots 0 \ldots)$ and $c_0(\ldots 1_k \ldots; \ldots 1_l \ldots)$ designate the functionals of the field-free state in vacuum and respectively in the case ($N_k^+ = 1$, $N_l^- = 1$, all other $N = 0$). If we now form the expectation value of the operator $\rho(\vec{\zeta})$ for the state $c(\ldots 0 \ldots; \ldots 0 \ldots)$, we obtain

$$\rho(\vec{\zeta}) = -\frac{1}{2}\Phi_0(\vec{\zeta})\sum_k \frac{(E_k - E_l)^2}{E_kE_l(E_k + E_l)},\ \vec{l} = \vec{k} + \vec{\zeta}.$$

The sum over k diverges logarithmically, as can easily be seen. After integration over the directions of $\vec{k}$, one in fact obtains:

$$\rho(\vec{\zeta}) = -\frac{e}{12hc\pi^2\zeta^2}\Phi_0(\vec{\zeta})\int_{k_0}^{\infty}\frac{\mathrm{d}|k|}{|k|} + \text{finite terms}.$$

If one returns to coordinate space, this yields:

$$\begin{aligned} \bar{\rho}(x) &= K\Delta\Phi_0 + \text{finite terms} \\ K &= \frac{1}{12\pi^2}\frac{e}{hc}\int\frac{\mathrm{d}|k|}{|k|}. \end{aligned}$$

The induced charge density has the opposite sign to the external density $\rho_0 = -(1/4\pi)\Delta\Phi_0$ and is proportional to it, with the diverging proportionality factor $4\pi K$. This would have as its consequence that every external charge would be completely compensated for by the induced charge. This result agrees completely with the result calculated by Dirac[16] on the basis of his hole theory. Even the factor K of the diverging term is the same.

Zürich, Physical Institute of the E. T. H.

[16] P. A. M. Dirac, Solvay Report 1933.

8 The electrodynamics of the vacuum based on the quantum theory of the electron

V. WEISSKOPF

Mathematisk-Fysiske Meddelelser det Kgl. Danske Videnskabernes Selskab, 14(6): 3–39 (1936).

[Weisskopf's paper was originally printed with the following English-language abstract.]
This paper deals with the modifications introduced into the electrodynamics of the vacuum by Dirac's theory of the positron. The behaviour of the vacuum can be described unambiguously by assuming the existence of an infinite number of electrons occupying the negative energy states, provided that certain well defined effects of these electrons are omitted, but only those to which it is obvious that no physical meaning can be ascribed.
The results are identical with these [*sic*] of Heisenberg's and Dirac's mathematical method of obtaining finite expressions in positron theory. A simple method is given of calculating the polarizability of the vacuum for slowly varying fields.

I

One of the most important results in the recent development of electron theory is the possibility of converting electromagnetic field energy into matter. For example, a light quantum, in the presence of other electromagnetic fields, can be absorbed in empty space and can be converted into matter. Here, a pair of electrons with opposite charge is created.

If the field where absorption proceeds is static, conservation of energy requires that the absorbed light quantum provides the total energy necessary to create the electron pair. The frequency of the light quantum thus must satisfy the relation $h\nu = 2mc^2 + \varepsilon_1 + \varepsilon_2$, where mc^2 is the rest energy of an electron and ε_1 and ε_2 are the residual energies of the two electrons. We encounter this case, for example, in the creation of an electron pair by a γ-quantum in the Coulomb field of an atomic nucleus.

Absorption can also take place in fields which come from other light quanta, where the latter can contribute to the energy of the electron pair. Consequently, in this case, the energy $2mc^2 + \varepsilon_1 + \varepsilon_2$ of the two electrons must be equal to the sum of all the light quanta absorbed in this process.

The phenomenon of the absorption of light in a vacuum represents an essential deviation from Maxwell's electrodynamics. Indeed, the vaccum clearly should be penetrable for a light quantum, independent of the fields prevailing there, since various fields can superpose independently according to Maxwell's equations, due to the linearity of these equations.

Without discussing the special theory more closely, it is already clear that, even in fields which do not have the energy necessary to create an electron pair, deviations from Maxwell's electrodynamics must occur: If higher frequency light can be absorbed in electromagnetic fields, one will expect a scattering or reflection of light beams whose frequency is not sufficient for pair creation. This is analogous to the scattering of light at an atom whose smallest absorption frequency is larger than that of the light. When passing through electromagnetic fields, light will behave as if the vacuum, under the action of the fields, were to acquire a dielectric constant different from unity.

To be able to represent these phenomena, theory must ascribe to empty space certain properties which cause the above-mentioned deviations from Maxwell's electrodynamics. In fact, the relativistic wave equation of the electron does indeed lead to such conclusions if one draws upon states of negative kinetic energy to describe the vacuum, states such as follow from the Dirac wave equation.

The basic assumption of the Dirac theory of the positron consists in the physical behavior of the vacuum in a certain sense being describable by the behavior of an infinite number of electrons – the vacuum electrons – which are situated in states of negative kinetic energy and hold all these states occupied. Naturally, agreement cannot be complete since the vacuum electrons have infinite charge and current densities which surely may not have any physical significance. But it appears, for instance, that pair creation (and its inverse process) are well reproduced by a jump of a vacuum electron into a state of positive energy under the influence of electromagnetic fields. The electron thus appears as a real electron, while the vacuum is depleted by a negative electron, which must express itself by the appearance of a positive electron. The calculation of pair creation and annihilation which starts from this picture shows good agreement with experiment.

In the calculation of most other effects which follow from the positron theory, one always encounters the problem to what extent the behavior of the vacuum electrons actually is to be regarded as the behavior of the vacuum. This problem is aggravated still more by the circumstance that the charge-, current-, and energy-densities of the vacuum electrons are infinite, so that what is generally involved is the separation, in a unique fashion, of a finite portion from an infinite sum, and to ascribe reality to this finite portion. This problem was solved by Dirac and Heisenberg by specifying a self-consistent method for determining the physically meaningful part of the action of the vacuum electrons. Below it will be shown that this specification is essentially free from any arbitrariness, since, in consistent fashion, it assumes only the following properties of the vacuum electrons as being physically meaningless:

I {
(1) The energy of the vacuum electrons in field-free space;
(2) the charge- and current-densities of the vacuum electrons in field-free space;
(3) a field-independent electric and magnetic polarizability of the vacuum, which is constant in both space and time.

These quantities[1] refer only to the field-free vacuum. It may be regarded as obvious that these quantities cannot have a physical meaning. All three quantities prove to be divergent sums after summing the contributions of all the vacuum electrons. It should

[1] To regard the assumptions (1) or (2) or (3) as meaningless will be quoted below by I_1, I_2 and I_3, respectively.

also be added that the polarizability could in no way be observed, but would only multiply all charges and field strengths by a constant factor.

In the next section, we shall calculate the physical properties of the vacuum in the presence of fields on the basis of these assumptions, where the fields vary slowly in time and in space. By this we understand such fields F which change only slightly along distances of length h/mc and during times of h/mc^2.[2] These fields thus satisfy the conditions

$$\frac{h}{mc}|\text{grad } F| \ll |F|, \frac{h}{mc^2}\left|\frac{\partial F}{\partial t}\right| \ll |F|. \tag{1}$$

In the presence of such fields, no pairs will generally be created, since the light quanta that are present have too little energy. The extreme cases, in which the radiation density is so high as to permit the collaboration of very many quanta, or in which electrostatic fields with potential differences in excess of $2mc^2$ are present (in this case, pairs would arise on the basis of Klein's paradox) are to be excluded from this consideration. Under these circumstances, the electromagnetic properties of the vacuum can be represented by a field-dependent electric and magnetic polarizability of empty space which leads, for example, to the refraction of light in electric fields or to the scattering of light from light. The dielectric and permeability tensors of the vacuum then have approximately the following form for weaker field strengths: ($\vec{E}$, $\vec{H}$, $\vec{D}$, $\vec{B}$ are here the four electromagnetic field variables[3])

$$\left.\begin{aligned}
D_i &= \sum_k \varepsilon_{ik} E_k, \; H_i = \sum_k \mu_{ik} B_k \\
\varepsilon_{ik} &= \delta_{ik} + \frac{e^4 h}{45\pi m^4 c^7}[2(E^2 - B^2)\delta_{ik} + 7B_i B_k] \\
\mu_{ik} &= \delta_{ik} + \frac{e^4 h}{45\pi m^4 c^7}[2(E^2 - B^2)\delta_{ik} + 7E_i E_k] \\
\delta_{ik} &= \begin{cases} 1, & i = k \\ 0, & i \neq k \end{cases}
\end{aligned}\right\} \tag{2}$$

The calculation of these phenomena has already been performed by Euler and Kockel[3] and by Heisenberg and Euler.[4] However, much simpler methods will be used in the next section. Furthermore, the properties of the vacuum will be calculated on the basis of the scalar relativistic wave equation of the electron due to Klein and Gordon. According to Pauli and Weisskopf,[5] this wave equation yields the existence of positive and negative particles, and their creation and annihilation by electromagnetic fields, without any special supplementary assumption. However, these particles have no spin and obey Bose statistics. For this reason, this theory cannot be applied to real electrons. However, it is noteworthy that this theory also leads to properties of the vacuum to which one cannot ascribe any physical significance. For instance, one thus likewise obtains an infinite field-independent polarizability of the vacuum, which is constant in

[2] h is Planck's constant divided by 2π.
[3] Below, arrows will be placed above vector quantities only where confusion is possible.
[3] H. Euler and B. Kockel, *Naturwiss.* **23**, 246, 1935; H. Euler, *Ann. d. Phys.* V **26**, 398.
[4] W. Heisenberg and H. Euler, *ZS. f. Phys.* **38**, 714, 1936.
[5] W. Pauli and V. Weisskopf, *Helv. Phys. Acta* **7**, 710, 1934.

space and time. After omitting the appropriate terms, one arrives at results similar to those of Dirac's positron theory. The physical properties of the vacuum in this theory arise from the 'zero-point energy' of matter which depends on external field strengths even in the absence of particles, and which thus yields a supplementary term to the pure Maxwell field energy.

In [this] section, we treat the conclusions from the Dirac positron theory for the case of general external fields, and we show that one always arrives at finite and unique results on the basis of the above-mentioned three assumptions concerning the action of the vacuum electrons. The Heisenberg subtraction methods proved to be identical to these three assumptions and thus appear significantly less arbitrary than has hitherto been assumed in the literature. All the following calculations did not take into account explicitly the mutual interactions of the vacuum electrons but consider exclusively each individual vacuum electron by itself under the influence of an existing field. With this method, however, the mutual effects are not completely neglected, since the external field cannot be separated from the field that is generated by the vacuum electrons themselves, so that the field entering the calculation contains implicitly in part the effects of the other vacuum electrons. This procedure is analogous to the Hartree calculation of electron orbits of an atom in the field that is modified by the electrons themselves. To calculate the interaction explicitly, one would have to use quantum electrodynamics, i.e. one would have to quantize the wave fields. As is well known, this already leads to divergences, even without assuming infinitely many vacuum electrons, and will not be touched upon in more detail in the following discussion.

II

This section will treat the electrodynamics of the vacuum for fields which satisfy the conditions (1). The field conditions are specified by specifying the energy density U as a function of the field strengths. These are determined from the energy density $\widetilde{U}$ of the vacuum electrons, which are supposed to be important for the behavior of the vacuum.

It is advantageous to fall back on the Lagrange function L of the electromagnetic field, since this function is already essentially specified by the requirement of relativistic invariance. The following relations exist between the Lagrange function L and the energy density U:

$$U = \sum_i E_i \frac{\partial L}{\partial E_i} - L. \tag{3}$$

In Maxwell's electrodynamics, the following holds:

$$L = \frac{1}{8\pi}(E^2 - B^2), \; U = \frac{1}{8\pi}(E^2 + B^2).$$

The additions to this Lagrange function must be relativistically invariant just like this function itself. As long as we limit ourselves only to slowly varying fields (condition (1)), these additions will depend only on the magnitudes of the field strengths and not on their derivatives. Consequently, they can only be functions of the invariants $(E^2 - B^2)$ and $(EB)^2$. If we expand these conditions in powers of the field strengths up to sixth order, we obtain:

$$L = \frac{1}{8\pi}(E^2 - B^2) + L'$$
$$L' = \alpha(E^2 - B^2)^2 + \beta(EB)^2 + \xi(E^2 - B^2)^3 + \zeta(E^2 - B^2)(EB)^2 + \cdots$$

and consequently according to (3)

$$\left.\begin{aligned} U &= \frac{1}{8\pi}(E^2 + B^2) + U' \\ U' &= \alpha(E^2 - B^2)(3E^2 + B^2) + \beta(EB)^2 \\ &\quad + \xi(E^2 - B^2)^2(5E^2 + B^2) + \zeta(EB)^2(3E^2 - B^2) + \cdots. \end{aligned}\right\} \quad (4)$$

The addition to the energy density is thus essentially specified by the invariance properties; consequently, below it will only be necessary to determine the constants which occur, namely α, β, ξ, ζ, ... These formulations are already based on the special assumption that U' does not contain second order terms in the field strengths but only higher order terms. This is equivalent to the vacuum having no polarizability independent of the fields.

The calculations of Euler and Kockel and of Heisenberg and Euler yield the following values for the four constants:

$$\alpha = \frac{1}{360\pi^2}\frac{e^4 h}{m^4 c^7}, \beta = 7\alpha, \xi = \frac{1}{630\pi^2}\frac{e^6 h^3}{m^8 c^{13}}, \zeta = \frac{13}{2}\xi.$$

The dielectric and permeability tensors given in (2) are derived from the relations:

$$D_i = 4\pi\frac{\partial L}{\partial E_i}, \; H_i = -4\pi\frac{\partial L}{\partial B_i}.$$

We shall derive these results below in a much simpler fashion.

The addition U' to Maxwell's energy density of the vacuum shall be determined by the addition $\widetilde{U}'$, which the vacuum electrons contribute. When electrons are present in the states $\psi_1, \psi_2, \ldots, \psi_i, \ldots$, the energy density is given by[6]

$$U = \frac{1}{8\pi}(E^2 + B^2) + \widetilde{U}'$$
$$\widetilde{U}' = \sum_i \left\{\psi_i^*, \left[\left(\vec{\alpha}, \frac{hc}{i}\text{grad} + e\vec{A}\right) + \beta mc^2\right]\psi_i\right\}$$

where $\vec{\alpha}$, β are the Dirac matrices and where $\vec{A}$ is the vector potential. The addition $\widetilde{U}'$ to the Maxwell density is thus not equal to the total material energy density U_{mat}

$$U_{\text{mat}} = ih\sum_i\left\{\psi_i^*, \frac{\partial}{\partial t}\psi_i\right\} \quad (5)$$

but

$$\widetilde{U}' = U_{\text{mat}} - \sum_i\{\psi_i^*, eV\psi_i\}, \quad (6)$$

where V is the scalar potential. One can regard $\widetilde{U}'$ as the kinetic energy density. The

[6] Two eigenfunctions ψ and φ in curly brackets $\{\psi, \varphi\}$ here and below designate the inner product of the two spinors ψ and φ: $\{\psi, \varphi\} = \sum_k \psi^k \varphi^k$, where k is the spin index.

total material energy density U_{mat} can easily be calculated, as we shall see; the second term of (6) – the potential energy density – is derived from U_{mat} in the following fashion. If one thinks of the scalar potential as proportional to the constant factor λ, one has[7]

$$\lambda \int \sum_i \{\psi_i^*, eV\psi_i\}\, \mathrm{d}\tau = \lambda \frac{\partial}{\partial \lambda} \int U_{\text{mat}}\, \mathrm{d}\tau. \tag{7}$$

Here, the integrations extend over all space. In the limiting case of constant fields, which we shall consider here because of conditions (1), we can regard the field strength E itself as the constant factor λ and we can furthermore transfer relation (7) also to the energy densities. We then obtain for the kinetic energy density

$$\tilde{U}' = U_{\text{mat}} - E \frac{\partial U_{\text{mat}}}{\partial E}. \tag{7a}$$

If one compares this with (3), one sees that the same relationship exists between the material and kinetic energy densities as between $-L$ and U. U_{mat} can thus be set equal to the addition to the Lagrange function, which here is caused by the vacuum electrons:

$$U_{\text{mat}} = -\tilde{L}'. \tag{8}$$

Since the form of U' is essentially specified by relativistic invariance requirements, it is sufficient to determine U' for a special field. We choose a homogeneous magnetic field $B = B(B_x, 0, 0,)$ and an electrostatic field that is parallel to this and that is periodic in space. The potential of this electrostatic field is given by

$$V = V_0 \exp(\mathrm{i}gx/h) + V_0^* \exp(-\mathrm{i}gx/h). \tag{9}$$

We then compare this result with the general form (4) and from this we will determine the coefficients of this form.

In contrast to this, Heisenberg and Euler chose a constant electric field. This causes difficulties as a consequence of Klein's paradox: Every homogeneous electric field, no matter how weak, generates electron pairs if it extends over all space. The electron occupation of energy states is then not exactly stationary. In the present calculation, the periodicity prevents potential differences above $2mc^2$ from occurring, so that no pair creations take place.

With full occupation of all negative energy states, the material energy density is given by

$$U_{\text{mat}} = \sum_i W_i \{\psi_i^*, \psi_i\}. \tag{10}$$

W_i is the energy associated with the eigenfunction ψ_i, and the summation extends over all negative states. The sum is naturally infinite. Which finite part of this sum is physically significant will appear unambiguously from the explicit expression for U_{mat}.

The ψ_i obey the wave equation:

$$\left\{\frac{\mathrm{i}h}{c}\frac{\partial}{\partial t} - \frac{eV}{c} + \alpha_x \mathrm{i}h \frac{\partial}{\partial x} + K\right\}\psi = 0, \tag{11}$$

[7] The proof is as follows: If the energy operator H depends on a parameter λ, the diagonal element H_{ii} of the energy operator will change as follows during an infinitesimal adiabatic change of λ by $\mathrm{d}\lambda$: $\mathrm{d}H_{ii} = (\partial H/\partial \lambda)_{ii}\, \mathrm{d}\lambda$. If we now set $H = H_0 + \lambda eV$, one then has $\lambda(eV)_{ii} = \lambda(\partial H_{ii}/\partial \lambda)$.

$$K = \alpha_y ih \frac{\partial}{\partial y} + \alpha_z \left[ih \frac{\partial}{\partial z} - \frac{e}{c}|B|y\right] - \beta mc. \tag{12}$$

For the time being we follow the calculation of Heisenberg and Euler, *loc. cit.*, where we make only unimportant changes in the meaning of the variables.

As a solution we set:

$$\psi_i = \frac{1}{(2\pi h)^{1/2}} \exp\left(\frac{i}{h} p_z z\right) \cdot u(y) X(x). \tag{13}$$

The operator K, when applied twice to ψ, yields:

$$K^2\psi = \left[-h^2 \frac{\partial^2}{\partial y^2} - i\alpha_y\alpha_z \frac{eh}{c}|B| + \left(p_z + \frac{e}{c}|B|y\right)^2 + m^2c^2\right]\psi.$$

We now set

$$\eta = \left(y + \frac{2p_z h}{b}\right)\left(\frac{b}{2h^2}\right)^{1/2}, \quad b = \frac{2eh}{c}|B|.$$

Here, b is the measure of the magnetic field. The reason for introducing η is that K^2 then obtains the form of an oscillator Hamiltonian function. We thus set

$$u(y) = \widetilde{H}_n(\eta)\left(\frac{b}{2h^2}\right)^{1/4},$$

where $\widetilde{H}_n(\eta)$ is the nth oscillator eigenfunction normalized to 1. There then holds $\int |u(y)|^2 \, dy = 1$ and

$$K^2\psi = \left\{m^2c^2 + b\left(n + \frac{1 - \sigma_x}{2}\right)\right\}\psi, \quad \sigma_x = i\alpha_y\alpha_z. \tag{14}$$

One can now choose a representation of the four-component ψ which is diagonal in σ_x:

$$\sigma_x = \begin{pmatrix} 1 & 0 & 0 & 0 \\ 0 & 1 & 0 & 0 \\ 0 & 0 & -1 & 0 \\ 0 & 0 & 0 & -1 \end{pmatrix}.$$

A positive spin in the x-direction then corresponds to the first two components of ψ, and a negative spin corresponds to the two others. With this choice, the wave equation (11) is decomposed into two separate systems of equations for the two pairs of components with the same spin, so that we obtain two wave equations with two-row matrices. The operator K can then be written in the form $K = \gamma|K|$, where γ is a two-row matrix which fulfils the condition $\gamma^2 = 1$ and where $|K|$ is the ordinary number

$$|K| = \left[m^2c^2 + b\left(n + \frac{1 - \sigma_x}{2}\right)\right]^{1/2}, \tag{15}$$

which depends on the value σ_x of the spin. The matrix α_x, which appears in the wave equation, likewise has two rows and anticommutes with γ: $\alpha_x\gamma + \gamma\alpha_x = 0$, since α_x, according to (12), also anticommutes with K. The two wave equations can then be written in the form:

$$\left\{\frac{ih}{c}\frac{\partial}{\partial t} + \alpha_x ih \frac{\partial}{\partial x} - \frac{e}{c}V + \gamma|K|\right\}\psi = 0 \tag{16}$$

where α_x and γ are two-row matrices which refer only to one component pair of the same spin. The difference in the wave equation for the two spin directions lies only in the different value of $|K|$. Since the dependence of ψ on the variables y and z has already been specified by (13), (16) represents a wave equation for the function $X(x)$ alone. Up to now, the course of the calculation is essentially identical with that of Heisenberg and Euler.

Now we treat the case $V = 0$. The eigenvalues and the normalized eigenfunctions for (16) then are

$$X_n^{(\pm)}(p_x) = a^{(\pm)}(p_x)(2\pi h)^{-1/2}\exp\left(\frac{ip_x x}{h}\right)\exp\left[\frac{iW_n^{\pm}(p_x)}{h}t\right], \tag{17}$$

$$W_n^{(\pm)}(p_x) = \pm c(p_x^2 + |K|^2)^{1/2} = \pm c\left[p_x^2 + m^2c^2 + b\left(n + \frac{1-\sigma_x}{2}\right)\right]. \tag{18}$$

The upper index (+) or (−) distinguishes the states of positive and negative energy. $\alpha^{\pm}(p)$ is a normalized two-component 'spinor'. Equation (16) and its solutions (17), (18) represent a one-dimensional analog to the Dirac equation, in which $\gamma|K|\psi$ replaces the mass term $\beta mc\cdot\psi$. Each momentum p_x has associated with it a positive and a negative energy value. (The two other energy values are given by the wave equations with opposite spin.)

If we now insert these quantities into the energy density (10), we obtain

$$U_{\mathrm{mat}} = \sum_{\sigma=-1}^{+1}\sum_{n=0}^{\infty}\iint\frac{\mathrm{d}p_x\,\mathrm{d}p_z}{2\pi h}W_n^-(p_x)|\tilde{H}(\eta)|^2\left(\frac{b}{2h^2}\right)^{1/2}|X_n^{(-)}(p_x)|^2.$$

Because $\mathrm{d}p_z = [(b/2)]^{1/2}\,\mathrm{d}\eta$ and $|X_n^{(-)}(p)|^2 = 1/2\pi h$, the integration over p_x yields

$$U_{\mathrm{mat}} = \frac{b}{8\pi^2h^3}\sum_{\sigma=-1}^{+1}\sum_{n=0}^{\infty}\int_{+\infty}^{-\infty}\mathrm{d}pW_n^-(p). \tag{19}$$

From here on, we shall write p instead of p_x.

To perform the summation, we form

$$\sum_{\sigma=-1}^{+1}\sum_{n=0}^{\infty}W_n^- = W_0^- + 2\sum_{n=1}^{\infty}W_n^-.$$

We now use the Euler sum formula for a function $F(x)$:

$$\frac{1}{2}F(a) + \sum_{r=1}^{N}F(a+rb) + \frac{1}{2}F(a+Nb)$$
$$= \frac{1}{b}\left[\int_a^{a+Nb}F(x)\,\mathrm{d}x - \sum_{m=1}^{\infty}(-)^m\frac{B_m}{(2m)!}b^{2m}\{F^{2m-1)}(a+Nb) - F^{(2m-1)}(a)\}\right].$$

B_m is the mth Bernoulli number. $F^{(m)}(x)$ is the mth derivative of $F(x)$. If we apply this to (19), we obtain

$$U_{\mathrm{mat}} = \frac{1}{4\pi^2h^3}\int\mathrm{d}p\left[\int_0^{\infty}F(x)\,\mathrm{d}x + \sum_{m=1}^{\infty}b^{2m}\frac{B_m}{(2m)!}(-)^mF^{(2m-1)}(0)\right], \tag{20}$$

$$F(x) = -c(p^2 + m^2c^2 + x)^{1/2}.$$

In the special case of a pure magnetic field, one can set U_{mat} equal to $\widetilde{U}'$ in accordance with (7a). This expression already represents the energy density, in an expansion by powers of the magnetic field strength B. Now it is very easy to determine that part of the contribution U' of the vacuum electrons which is supposed to be decisive for the real vacuum: The term independent of B represents the energy density of the field-free vacuum and is a divergent integral; since the energy density for the field-free vacuum must vanish, this expression cannot have real significance. Furthermore, the terms with b^2 (which incidentally also diverge) must be omitted, since the energy density should not have terms of second order in the field strengths. The omission of these terms is based on the assumption that the polarizability of the vacuum tends to zero at vanishing fields. It should be emphasized that the subtractions occurring here are based exclusively on trivial assumptions concerning the field-free vacuum.

We thus obtain for the addition to the Maxwell energy density:

$$U' = -\frac{c}{4\pi^2 h^3} \sum_{m=2}^{\infty} \frac{B_m(-)^m}{(2m)!} b^{2m} \frac{1{\cdot}3 \dots (4m-5)}{2^{2m-1}} \times \int_{-\infty}^{+\infty} \frac{\mathrm{d}p}{(p^2+m^2c^2)(4m-3)/2}. \tag{21}$$

This power series can easily be represented by the power-series expansion of the hyperbolic cotangent, Ctg. One obtains:

$$U' = \frac{1}{8\pi^2} mc^2 \left(\frac{mc}{h}\right)^3 \int_0^{\infty} \frac{\mathrm{d}\eta}{\eta^3} \mathrm{e}^{-\eta} \left\{ \eta \boldsymbol{B} \operatorname{Ctg} \eta \boldsymbol{B} - 1 - \frac{\eta^2}{3} \boldsymbol{B}^2 \right\}$$

where $\boldsymbol{B}$ is the magnetic field strength measured in units of the critical field strength m^2c^3/eh:

$$\boldsymbol{B} = \frac{eh}{m^2c^3} B.$$

The first and second terms of the expansion yield:

$$U' = -\frac{1}{360\pi^2} \frac{e^4 h}{m^4 c^7} B^4 + \frac{1}{630\pi^2} \frac{e^6 h^3}{m^8 c^{13}} B^6 + \cdots$$

If we compare this with those terms of (4) which contain the magnetic field to the fourth and sixth power, we obtain:

$$\alpha = \frac{1}{360\pi^2} \frac{e^4 h}{m^4 c^7}, \quad \xi = \frac{1}{630\pi^2} \frac{e^6 h^3}{m^8 c^{13}}.$$

Now let us also consider the electric field. For this purpose, we solve the wave equation (16) for $X(x)$ with the Born approximation method. We expect that the parts of U_{mat} which are independent of the potential V will appear in the second approximation which is proportional to V^2. If we expand U_{mat} in powers of V: $U_{\text{mat}} = U_{\text{mat}}^{(0)} + U_{\text{mat}}^{(1)} + \cdots$ we obtain, according to (10):

$$U_{\text{mat}}^{(2)} = \sum_i W_i^{-(2)} (|\psi_i|^2)^{(0)} + \sum_i W_i^{-(0)} (|\psi_i|^2)^{(2)}. \tag{22}$$

$W_i^{-(k)}$, $(|\psi_i|^2)^{(k)}$ are the kth approximations in the corresponding expansion of W_i^- and $|\psi_i|^2$. It should be noted that $W_i^{-(1)}$ vanishes in the given electric field. It can easily be shown that

$$\int (|\psi|^2)^{(2)} \, dx \, dy \, dz = 0$$

so that the space average of U_{mat} is given only by the first term in (22):

$$\overline{U_{\text{mat}}^{(2)}} = \sum_i W_i^{-(2)} (|\psi_i|^2)^{(0)}.$$

$(|\psi_i|^2)^{(0)}$ was already calculated in the case of a pure magnetic field and we thus obtain, quite analogously to (19),

$$\overline{U_{\text{mat}}^{(2)}} = \frac{b}{8\pi^2 h^3} \sum_{\sigma=-1}^{+1} \sum_{n=0}^{\infty} \int_{-\infty}^{+\infty} dp W_n^{-(2)}(p).$$

The value of $W_n^{-(2)}$ can be calculated by means of the Born approximation method. With the eigenfunctions (17), one obtains:

$$\left. \begin{aligned} W_n^{-(2)}(p) = e^2 |V_0|^2 \Bigg[& \frac{|\{a^{(+)*}(p+g), a^{(-)}(p)\}|^2}{W_n^-(p) - W_n^+(p+g)} \\ & + \frac{|\{a^{(-)*}(p+g), a^{(-)}(p)\}|^2}{W_n^-(p) - W_n^-(p+g)} \Bigg] \\ & + \text{the same with } -g. \end{aligned} \right\} \tag{23}$$

The expression within curly brackets represents a scalar product of two two-component spinors. In the integration of (23) over p, the second terms in the square brackets are eliminated if the integration of the terms with $-g$ is performed with the variables $p' = p - g$:

$$\left. \begin{aligned} \int dp W_n^{-(2)} = e^2 |V_0|^2 \int dp & \frac{|\{a^{(+)*}(p+g), a^{(-)}(p)\}|^2}{W_n^{(-)}(p) - W_n^{(+)}(p+g)} \\ & + \text{the same with } -g. \end{aligned} \right\} \tag{24}$$

This procedure is generally in no way unique, since the integration of the second term in the square brackets of (23) leads to a divergent result, which, however, can be made finite by adding the corresponding term with $-g$. Alternatively, as in (24), it can be made to vanish depending on the manner in which the integration variables are chosen. This arbitrariness does not affect our calculation however, since, after performing the summation over n, we use only the terms proportional to b^2, b^4, etc., in which, on the basis of the Euler sum formula, there occur only derivatives of $W_n^{-(2)}(p)$ with respect to n. As one can easily convince oneself, these derivatives of the second term in the square brackets no longer diverge in the integration over p, so that the result of this integration is independent of the choice of integration variables.

From (24), one further obtains:

$$\int dp W_n^{-(2)} = -e^2 |V_0|^2 \frac{g^2}{4} \int dp \frac{|K|^2}{c(p^2 + |K|^2)^{5/2}}$$

where a series expansion in powers of g has already been performed and terms of order higher than the second have been omitted. This implies that one should neglect the derivatives of the field strengths on the basis of conditions (1). Just as in the previous calculation, $\overline{U^{(2)}_{\text{mat}}}$ is given by (20), if one sets:

$$F(x) = -e^2|V_0|^2 \frac{g^2}{4} \frac{m^2c^2 + x}{c(p^2 + m^2c^2 + x)^{5/2}}.$$

One then obtains, by first integrating over p:

$$\overline{U^{(2)}_{\text{mat}}} = -\frac{1}{4\pi^2 h^3 c} \frac{g^2}{3} e^2|V_0|^2 \left[\int_0^\infty \frac{\mathrm{d}x}{m^2c^2 + x} + \sum_{m=1}^{\infty} b^{2m} \frac{B_m(-)^m}{(2m)!} \left(\frac{\mathrm{d}^{2m-1}}{\mathrm{d}x^{2m-1}} \frac{1}{m^2c^2 + x}\right)_{x=0}\right].$$

Since this expression is quadratic in the electric field strengths, we obtain for the kinetic energy density, according to (7a),

$$\overline{\widetilde{U}'^{(2)}} = -\overline{U^{(2)}_{\text{mat}}}.$$

For reasons that have already been discussed, only the terms of fourth and higher order in the field strengths can have physical significance for the vacuum, so that the diverging integral is to be omitted. We now replace V_0 by the electric field strength E:

$$\overline{E^2} = 2\frac{g^2}{h^2}|V_0|^2,$$

where the bars signify space averages. For the first two terms we obtain:

$$U'^{(2)} = \frac{5}{360\pi^2} \frac{e^4 h}{m^4 c^7} E^2 B^2 - \frac{7}{2} \frac{1}{630\pi^2} \frac{e^6 h^3}{m^8 c^{13}} E^2 B^4 + \cdots \tag{25}$$

for the limiting case of slowly varying fields, where the space averages can be omitted.

If one compares (25) with the terms in (4) that are proportional to E^2B^2 and E^2B^4, one obtains the relations

$$\beta - 2\alpha = \frac{5}{360} \frac{e^4 h}{m^4 c^7}, \ 3\xi - \zeta = \frac{7}{2} \frac{1}{630\pi^2} \frac{e^6 h^3}{m^8 c^{13}},$$

and, using the already calculated values of α and β,

$$\beta = 7\alpha, \ \zeta = \frac{13}{2}\xi.$$

The expression for $U'^{(2)}$, which is exact in the magnetic field strengths, turns out to be

$$U'^{(2)} = \frac{1}{8\pi^2} mc^2 \left(\frac{mc}{h}\right)^3 \frac{1}{3} \boldsymbol{E}^2 \int_0^\infty \frac{\mathrm{d}\eta}{\eta} \mathrm{e}^{-\eta} \{\eta \boldsymbol{B} \operatorname{Ctg} \boldsymbol{B} - 1\},$$

where $\boldsymbol{E} = (eh/m^2c^3)E$.

The higher approximations in E can be determined easily up to a constant factor. We think of the kth approximation $\overline{W}_n^{(k)}(p)$ of the energy as being determined by the state specified by p and n; because of the wave equation (16), it will have the following form:

$$\overline{W}_n^{(k)}(p) = g^k e^k |V_0|^k \cdot G(c, h, |K|, p),$$

where G is a function in which only the specified quantities occur. Because of gauge invariance, $W^{(k)}$ must be of at least kth order in g. The higher powers of g are neglected. The energy density of kth order then becomes:

$$U_{\text{mat}}^{(k)} = \frac{-1}{4\pi^2 h^3} g^k e^k |V_0|^k \left[\int_0^\infty \mathrm{d}x \int_{-\infty}^{+\infty} G \, \mathrm{d}p \right.$$
$$\left. + \sum_{m=1}^{\infty} b^{2m} \frac{B_m}{(2m)!} (-)^m \left(\frac{\mathrm{d}^{2m-1}}{\mathrm{d}x^{2m-1}} \int_{-\infty}^{+\infty} G \, \mathrm{d}p \right)_{x=0} \right]. \tag{26}$$

The integral over G must have the dimension $(\text{energy})^{-(k-1)}$ $(\text{momentum})^{-(k-1)}$ and now may only depend on the quantities c, h, $|K|$. This is possible only in the form:

$$\int_{-\infty}^{+\infty} G \, \mathrm{d}p = f_k \frac{1}{c^{k-1}|K|^{2k-2}} = f_k \frac{1}{c^{k-1}(m^2c^2 + x)^{k-1}}$$

where f_k is a numerical factor.

If this is inserted into (26), $U_{\text{mat}}^{(k)}$ can be specified completely up to the factor f_k. The numerical factors f_k, however, are easily determined from the consideration that U_{mat}, according to (8), must be a relativistic invariant. Consequently U_{mat} can depend only on $E^2 - B^2$ and $(EB)^2$. Thus, for example, the coefficient of E^k may differ from the coefficient of B^k only by the factor $(-)^{k/2}$. The latter coefficient was already calculated and was given by (21). One then obtains

$$f_{2m} = \frac{2^{3m-2} B^m}{m(2m-1)}$$

and one can thus calculate the representation of L' which has been given by Heisenberg and Euler:[8]

$$L' = -\frac{1}{8\pi^2} mc^2 \left(\frac{mc}{h}\right)^3$$
$$\times \int_0^\infty \frac{\mathrm{d}\eta}{\eta^3} \mathrm{e}^{-\eta} \left\{ \eta \boldsymbol{B} \operatorname{Ctg} \eta \boldsymbol{B} \cdot \eta \boldsymbol{E} \operatorname{Ctg} \eta \boldsymbol{E} - 1 + \frac{\eta^2}{3} (\boldsymbol{E}^2 - \boldsymbol{B}^2) \right\}$$
$$\boldsymbol{E} = \frac{m^2 c^3}{eh} E, \; \boldsymbol{B} = \frac{m^2 c^3}{eh} B.$$

This expression has been calculated for parallel fields. To generalize it to arbitrary fields, one must write it as a function of the two invariants $E^2 - B^2$ and $(EB)^2$. According to Heisenberg and Euler, this is possible in simple fashion by means of the relation

$$\operatorname{Ctg} \alpha \operatorname{Ctg} \beta = -\mathrm{i} \frac{\cos(\beta^2 - \alpha^2 + 2\mathrm{i}\alpha\beta)^{1/2} + \text{conj.}}{\cos(\beta^2 - \alpha^2 + 2\mathrm{i}\alpha\beta)^{1/2} - \text{conj.}}$$

and one obtains

$$L' = \frac{1}{8\pi^2} \frac{e^2}{hc} \int_0^\infty \mathrm{e}^{-\eta} \frac{\mathrm{d}\eta}{\eta^3} \left\{ \mathrm{i}\eta^2 (EB) \frac{\cos \eta[(\boldsymbol{E}^2 - \boldsymbol{B}^2 + 2\mathrm{i}(\boldsymbol{EB})]^{1/2} + \text{conj.}}{\cos \eta[(\boldsymbol{E}^2 - \boldsymbol{B}^2 + 2\mathrm{i}(\boldsymbol{EB})]^{1/2} - \text{conj.}} \right.$$
$$\left. + \frac{m^4 c^6}{e^2 h^2} + \frac{\eta^2}{3} (B^2 - E^2) \right\}.$$

Because of the reality of the total expression, the latter depends actually only on $E^2 - B^2$ and $(EB)^2$.

[8] As regards the question of the convergence of this integral, we refer to the relevant remarks in the paper by Heisenberg and Euler, page 729.

In the scalar theory of the positron, the energy density and Lagrange function of the vacuum are calculated with the same mathematical means. In this theory, the vacuum energy density arises from the zero-point energy of the matter waves. According to Pauli and Weisskopf, *loc. cit.* (Formula (29)), the total energy is given by

$$E_{\text{mat}} = \sum_k W_k(N_k^+ + N_k^- + 1)$$

where W_k is the energy of the kth state, N_k^+ is the number of positrons, and where N_k^- is the number of electrons belonging to the state. In the empty vacuum, there remains a sum over all energies W_k, where the energy W_k of the state characterized by the momentum p and the quantum number n in the magnetic field B has the value:

$$W_n^{\text{scal}}(p, B) = c[p^2 + m^2c^2 + b(n + \tfrac{1}{2})]^{1/2}.$$

Summation over all states and division by the total volume leads to the energy density, which is easily obtained analogously to (19):

$$U_{\text{mat}} = \frac{b}{8\pi^2 h^3} \sum_{n=0}^{\infty} \int_{-\infty}^{+\infty} \mathrm{d}p W_n^{\text{scal}}(p, B).$$

The only difference compared to the previous calculation consists in the omission of the summation over the two spin directions. Now one easily verifies the following relation between the energy $W_n^{\text{scal}}(p, B)$ in the scalar electron theory and the energy $W_n^-(p, B)$ in the Dirac electron theory:

$$2\sum_{n=0}^{N} W_n^{\text{scal}}(p, B) = \sum_{\sigma=-1}^{+1} \sum_{n=0}^{N} W_n^-(p, B) - \sum_{\sigma=-1}^{+1} \sum_{n=0}^{2N} W_n^-(p, B/2).$$

Consequently, we can express the energy density in the scalar theory $\widetilde{U}'_{\text{scal}}$ by the energy density $\widetilde{U}'$ in the Dirac positron theory, in the following fashion:

$$2\widetilde{U}'_{\text{scal}}(B) = \widetilde{U}'(B) - 2\widetilde{U}'(B/2).$$

From this, one can see that, here too, the part that is independent of the field strengths and that is quadratic in the field strengths is infinite. The latter thus yields an infinite polarizability independent of the field strengths. To obtain a useful result for the field-free vacuum, one must again delete these two portions, and consequently one obtains the relation

$$\operatorname{Ctg}\beta - 2\operatorname{Ctg}\frac{\beta}{2} = -\frac{1}{\operatorname{Sin}\beta}:$$

$$U'_{\text{scal}} = -\frac{1}{16\pi^2} mc^2\left(\frac{mc}{h}\right)^3 \int_{(0)}^{\infty} \frac{\mathrm{d}\eta}{\eta^3} \mathrm{e}^{-\eta}\left\{\eta \boldsymbol{B} \frac{1}{\operatorname{Sin}\eta\boldsymbol{B}} - 1 + \frac{\eta^2}{6}\boldsymbol{B}^2\right\}.$$

Performing an analogous perturbation calculation in the electric field in similar fashion leads to an addition to the Lagrange function of the field which is closely related to the one derived from the Dirac positron theory:

$$L'_{\text{scal}} = -\frac{1}{16\pi^2}\frac{e^2}{hc}\int_0^{\infty}\frac{\mathrm{d}\eta}{\eta^3}\mathrm{e}^{-\eta}\left\{\frac{2\mathrm{i}\eta^2(\boldsymbol{EB})}{\cos\eta[(\boldsymbol{E}^2 - \boldsymbol{B}^2) + 2\mathrm{i}(\boldsymbol{EB})]^{1/2} - \text{conj.}} + \frac{m^4c^6}{e^2h^2} - \frac{\eta^2}{6}(\boldsymbol{B}^2 - \boldsymbol{E}^2)\right\}.$$

For the coefficients α, β defined in (4), one therefore obtains:

$$\alpha = \frac{7}{16}\frac{1}{360\pi^2}\frac{e^4 h}{m^4 c^7}, \beta = \frac{4}{7}\alpha.$$

Reference should here also be made to the following property of the Lagrange function of the vacuum. For very great field strengths E or B, the highest terms of the addition L' to the Maxwell Lagrange function have the following form in the Dirac theory of the positron:

$$L' \sim -\frac{e^2}{24\pi^2 hc} E^2 \lg \boldsymbol{E} \text{ or } L' \sim \frac{e^2}{24\pi^2 hc} B^2 \lg \boldsymbol{B}.$$

The ratio between this addition L' and the Maxwell Lagrange function $L_0 = (1/8\pi)(E^2 - B^2)$ is thus logarithmic in the field strengths for high values of the latter and is furthermore multiplied by the factor e^2/hc:

$$\frac{L'}{L_0} \sim \frac{-e^2}{3\pi hc} \lg \boldsymbol{E} \text{ or } \frac{L'}{L_0} \sim -\frac{e^2}{3\pi hc} \lg \boldsymbol{B}.$$

The nonlinearities of the field equations thus represent only small corrections even for field strengths which are much higher than the critical field strength m^2c^3/eh. As stated in the note by Euler and Kockel, *loc. cit.*, and in the paper by Euler, *loc. cit.*, the nonlinearity of the field equations that follows from the positron theory is related to the nonlinear field theory of Born and Infeld,[9] but this relationship is thus only external. In the latter theory, the Maxwell equations are already completely altered at the critical field strength $F_0 = m^2c^4/e^3$ at the edge of the electron, and the finite self-energy of a point charge is then obtained in this way. Here, on the other hand, the deviations from the Maxwell field equations for fields of the order F_0 are still very small and grow much too slowly to play a similar role in the self-energy problem. The extrapolation of the present calculations to fields at 'the edge of the electron', however, is not free of misgivings since there the conditions (1) are not fulfilled. However, it is not probable that a more precise consideration in this respect would yield a very different result.

III

This section will treat the influence of arbitrary fields on the vacuum. At first we limit ourselves to static fields. The stationary state of the electron, on the basis of the Dirac wave equation and its eigenvalues, will generally be divisible into two groups, which have arisen from the positive and respectively negative energy levels of the free electrons when the static field is turned on adiabatically. This applies, for example, to the Coulomb field of an atomic nucleus and to all static fields that occur in nature.

However, one can also specify static fields such that a division of this kind will fail since, as a consequence of the field, transitions occur from negative to positive states. A well-known example of this is a potential step of height greater than $2mc^2$. These exceptional cases cannot be treated in the stationary fashion and must be regarded as a

[9] M. Born and L. Infeld, *Proc. Roy. Soc.* **143**, 410, 1933.

time-dependent field which is switched on at a given time. This is already necessary, all the more so because such fields cannot be maintained stationary due to continuous pair creation.

However, in the case where the eigenvalue spectrum can be unambiguously divided into the two groups, the energy density U and the current- and charge-densities $\vec{i}$, ρ of the vacuum electrons can be calculated according to the formulas

$$\left.\begin{aligned} U &= ih\sum_i\left\{\psi_i^*, \frac{\partial}{\partial_l}\psi_i\right\} \\ \rho &= e\sum_i\{\psi_i^*, \psi_i\} \\ \vec{i} &= e\sum_i\{\psi_i^*, c\vec{\alpha}\psi_i\} \end{aligned}\right\} \tag{29}$$

where the sum extends over the states which correspond to the negative energy states of the free electron. The sums as written will diverge. However, if the physically meaningless parts are separated out, we obtain convergent expressions.

In order to specify these parts, on the basis of the assumptions I, we expand the summands of expressions (29) in powers of the external field strength, in such a fashion that we think of the latter as being multiplied by a factor λ and then expand in powers of this factor. This procedure is identical with a successive perturbation calculation which starts from the free electrons as the zeroth approximation.

The assumptions I_1, I_2 demand clearly the vanishing of terms that are independent of λ, which consist of contributions from free vacuum electrons that are independent of the field. If we provisionally consider here only those free vacuum electrons whose momentum $|p| < P$, we obtain for this the following contributions:[10]

$$\left.\begin{aligned} U_0 &= -\frac{1}{4\pi^3h^3}\int_{|p|<P} \mathrm{d}\vec{p}\, c(p^2+m^2c^2)^{1/2}, \\ \rho_0 &= \frac{e}{4\pi^3h^3}\int_{|p|<P} \mathrm{d}\vec{p}, \\ \vec{i}_0 &= \frac{e}{4\pi^3h^3}\int_{|p|<P} \mathrm{d}\vec{p}\,\frac{c\vec{p}}{(p^2+m^2c^2)^{1/2}}. \end{aligned}\right\} \tag{30}$$

The contributions of all electrons – $P \to \infty$ – naturally diverge.

By separating the terms that are independent of λ, however, the assumption I_2 is not yet completely fulfilled. The charge- and current-densities ρ_0, $\vec{i}_0$ of the field-free vacuum electrons are in fact also expressed by the fact that, in the presence of potentials V, $\vec{A}$, they yield the additions $\rho_0 V$ and $(\vec{i}_0\vec{A})$ to the energy density, which must likewise be separated. These additions occur since the energy and the momentum of vacuum electrons, which are still unaffected by the fields, are changed, in the presence of potentials, by the amounts eV and $(e/c)\vec{A}$, respectively.

The assumptions I_1 and I_2 are therefore fulfilled completely only if the permitted contributions (30) are modified in the following fashion:

[10] The current associated with the momentum $\vec{p}$ is $ec\vec{p}/(p^2 + m^2c^2)$ and the number of states in the interval $\mathrm{d}\vec{p}$ is: $\mathrm{d}p/4\pi^3h^3$.

$$\left.\begin{aligned}
U_0' &= -\frac{1}{4\pi^3 h^3}\int_{|p|<P} \mathrm{d}\vec{p}\left[c\left[\left(p+\frac{e}{c}A\right)^2+m^2c^2\right]^{1/2}+eV\right]\\
\rho_0' &= \frac{e}{4\pi^3 h^3}\int_{|p|<P}\mathrm{d}\vec{p}\\
\vec{i}_0' &= \frac{e}{4\pi^3 h^2}\int_{|p|<P}\mathrm{d}\vec{p}\,-\frac{c\left(\vec{p}+\frac{e}{c}\vec{A}\right)}{\left[\left(p+\frac{e}{c}A\right)^2+m^2c^2\right]^{1/2}}.
\end{aligned}\right\} \tag{31}$$

Here, too, only the portion coming from free vacuum electrons with momenta $|p| < P$ has again been written down. Now the condition I_3 must still be fulfilled. In this connection, we note that a constant field-independent polarizability leads to terms in the energy density $U(x)$ which are proportional to the squares $E^2(x)$ and $B^2(x)$ of the field strengths at the point x. Likewise, it leads to current- and charge-densities which are proportional to the first derivatives of the fields, due to the relations

$$i = \operatorname{rot} M + \frac{\mathrm{d}P}{\mathrm{d}t},$$

$$\rho = \operatorname{div} P,$$

where M and P are the magnetic and electric polarizations which, in the case of a constant polarizability, are proportional to the fields. To fulfil the assumption I_3, the terms proportional to E^2 and B^2 must therefore vanish in the energy density U of the vacuum electrons and the terms proportional to the first derivatives must be omitted in the current- and charge-densities. It is more practical not to specify specifically the form of these terms, but to recognize these terms by their properties during the course of the calculation.

As an explanation, we calculate the charge- and current-densities of the vacuum under the influence of an electric potential

$$V = V_0 \exp\left(\frac{\mathrm{i}(\vec{g}\cdot\vec{r})}{h}\right) + V_0^* \exp\left(-\frac{\mathrm{i}(\vec{g}\cdot\vec{r})}{h}\right) \tag{32}$$

and a magnetic potential

$$\vec{A} = \vec{A}_0 \exp\left(\frac{\mathrm{i}(\vec{g}\cdot\vec{r})}{h}\right) + \vec{A}_0^* \exp\left(\frac{\mathrm{i}(\vec{g}\cdot\vec{r})}{h}\right), (\vec{A}_0, \vec{g}) = 0 \tag{33}$$

by means of perturbation theory. These calculations have already been performed by Heisenberg[11] and much more generally by Serber[12] and Pauli and Rose,[13] and will serve here only as an illustration of our physical interpretation of the subtraction terms. The charge density ρ is given up to first order by $\rho = \rho^{(0)} + \rho^{(1)}$

$$\rho^{(1)} = e\sum_i\sum_k \frac{H_{ki}\{\psi_i^*, \psi_k\}}{W_i - W_k} + \text{conj}.$$

where i sums over occupied states and k sums over unoccupied states, and H_{ik} is the

[11] Heisenberg, *Z. f. Phys.* **90**, 209, 1934.
[12] R. Serber, *Phys. Rev.* **48**, 49, 1935.
[13] W. Pauli and M. Rose, *Phys. Rev.* **49**, 462, 1936.

matrix element of the perturbation energy. If we introduce the potential (32) as a perturbation, we obtain $W(p) = c(p^2 + m^2c^2)^{1/2}$

$$\rho^{(1)} = \frac{e^2 V_0}{8\pi^3 h^3} \cdot \int d\vec{p} \left\{ \frac{W(p)W(p+g) - c^2(p, p+g) - m^2c^4}{W(p)W(p+g)[W(p) + W(p+g)]} \right.$$

$$\left. \text{(the same with } -g) \right\} \exp\left(\frac{i(\vec{g} \cdot \vec{r})}{h}\right) + \text{conj.}$$

If one expands this in powers of $\vec{g}$, one obtains

$$\rho^{(1)} = \frac{e^2 V_0}{8\pi^3 h^3} \int d\vec{p} \, \frac{c^2}{W^3(p)} \exp\left(\frac{i(\vec{g} \cdot \vec{r})}{h}\right)$$

$$\left\{ \frac{g^2}{2} - \frac{c^2(pg)^2}{W^2(p)} - \frac{c^2 g^4}{4W^2(p)} + \frac{25}{16} \frac{c^4(pg)^2 g^2}{W^4(p)} - \frac{21}{8} \frac{c^6(pg)^4}{W^6(p)} + \cdots \right\}$$

$$+ \text{conj.}$$

Now which part of this charge density has physical meaning? Because of the assumption I_2, ρ_0 must be deleted; in $\rho^{(1)}$, the terms with g^2 are proportional to the second derivative of V and thus to the first derivative of the field strengths, and must be deleted because of I_3. One should note that it is also only these terms which lead to divergences. The remainder yields finite integrals and, according to Heisenberg, is to be written in the form

$$\rho^{(1)} = \frac{1}{60\pi^2} \frac{e^2}{hc} \left(\frac{h}{mc}\right)^2 \Delta\Delta V. \tag{34}$$

The exact calculation of the first approximation was given by Serber and by Pauli and Rose, *loc. cit.*

As another example, we consider the current density $\vec{i}$ in the first approximation of the field (33)

$$\vec{i}^{(1)} = e \sum_{ik} \frac{H_{ki}\{\psi_i^*, \vec{\alpha}\psi_k\}}{W_i - W_k} + \text{conj.}$$

One obtains

$$\vec{i}^{(1)} = -e^2 \frac{c\vec{A}_0}{8\pi^3 h^3} \int d\vec{p} \exp\left(\frac{i(\vec{g} \cdot \vec{r})}{h}\right)$$

$$\left\{ \frac{W(p)W(p+g) + E^2(p) + c^2(pg) - 2c^2(n, p+g)(n, p)}{W(p+g)W(p)[W(p+g) + W(p)]} \right.$$

$$\left. \text{(the same with } -g) \right\} + \text{conj.},$$

where n is the unit vector in the direction $\vec{A}$. Expanding it in g, this yields

$$\begin{cases} \vec{i}^{(1)} = -\dfrac{e^2 \vec{c}A}{4\pi^3 h^3} \displaystyle\int d\vec{p} \, \frac{1}{W^3(p)} \\ \qquad \left\{ W^2(p) - c^2(np)^2 - \dfrac{c^2 g^2}{2} + \dfrac{3}{4} \dfrac{c^4(pg)^2}{W^2(p)} - \dfrac{5}{2} \dfrac{c^6(np)^2(pg)^2}{W^4(p)} \right. \\ \qquad + \dfrac{3}{4} \dfrac{c^4(np)^2 g^2}{W^2(p)} + \text{terms of fourth and higher order in } g. \end{cases} \tag{35}$$

Terms also occur here which are independent of g, and which thus are not gauge invariant. However, these are identical to the contributions $\vec{i}_0'$ from (31), contributions which are to be omitted: In fact, if one expands $\vec{i}_0'$ in terms of $\vec{A}$, one obtains

$$\vec{i}_0' = \frac{e}{4\pi^3 h^3} \int d\vec{p} \left\{ \frac{c p}{W(p)} + \frac{e\vec{A}}{W(p)} - \frac{c^2 p(e\vec{A}, \vec{p})}{W^3(p)} + \cdots \right\}$$

$$= \vec{i}_0 + e^2 \frac{c\vec{A}}{4\pi^3 h^3} \int d\vec{p} \frac{1}{W^3(p)} \{W^2(p) - c^2(np)^2 + \cdots\}.$$

The first order terms in $\vec{A}$ agree with the terms in (35) that are independent of g. The terms in (35) that are proportional to g^2 are likewise omitted, the remainder leads to the converging result that corresponds to (34):

$$\vec{i}^{(1)} = \frac{1}{60\pi^2} \frac{e^2}{hc} \left(\frac{h}{mc}\right)^2 \Delta\Delta\vec{A} + \text{higher derivatives}$$

These two examples should show that the contributions to be omitted are directly recognizable in the perturbation calculation and that the remaining contributions of the vacuum electrons, which are not affected by the assumptions I, no longer lead to divergences in the summation. The given examples indeed prove this only in a first approximation. But the considerations can easily be extended to higher approximations.

The treatment of time-dependent fields is not essentially different from the above procedure. It is necessary to let the time-dependent fields act from a time t_0, when the vacuum electrons were in a field-free state, or in such stationary states as can readily be decomposed into occupied and unoccupied states. The time change of these states, beginning at time t_0, can then be represented by means of a perturbation calculation in powers of the external fields. The expressions (31) and the terms following from the condition I_3 can then be separated, and the residue which remains no longer leads to divergences. The calculation of the charge- and current-densities of the vacuum for arbitrary time-dependent fields, to a first approximation, is found in Serber, *loc. cit.*, and in Pauli and Rose, *loc. cit.* The parts being subtracted are there formally taken from the Heisenberg paper. But they are completely identical to those which follow from the assumptions I.

How does the creation of pairs now express itself through time-dependent fields? In the calculation of the energy-, current-, and charge-densities, the pairs are not expressed directly. Pair creation appears only in a total energy which increases proportional to time, and which corresponds to the energy of the electrons being created. The charge- and current-densities are not directly influenced by pair creation, since positive and negative electrons are always created simultaneously. These influence the current- and charge-densities only by virtue of the fact that the external fields act differently on the resulting electrons depending on their charge.[14]

It is therefore more practical to calculate pair creation through external fields directly as a transition of a vacuum electron into a state of positive energy. The probability of

[14] Serber has calculated the current- and charge-densities in fields which create pairs. This density is thus to be ascribed to the resonant vibrations of the vacuum electrons and is not the 'created current- and charge densities'. The resonance denominators which appear come from the fact that this resonance is especially strong when the external frequency approaches the absorption frequency of the vacuum.

creating the electron pair is then identical to the increase of the intensity of the relevant eigenfunction of positive energy or respectively with the decrease of intensity of the corresponding eigenfunction of negative energy, due to the action of the time-dependent fields on states that already prevailed up to time t_0. This calculation has been performed by Bethe and Heitler,[15] and Hulme and Jaeger.[16]

The annihilation of pairs with radiation of light can be treated, like every other spontaneous radiation process, only by quantizing the wave fields, or by a reversal of the light absorption process in conformity with the correspondence principle.

In the presentation up to now, the portions of the vacuum electrons that are to be separated off were not given explicitly; only their form and their dependence on the external fields were determined. To represent them explicitly, one must choose a somewhat different procedure, since these parts do indeed contain divergent expressions. The density matrix introduced by Dirac is suitable for this. The density matrix was applied to this problem by Dirac and more especially by Heisenberg, *loc. cit*. The density matrix R is given by the following expression:

$$(x', k'|R|x''k'') = \sum_i \psi_i^*(x'k')\psi_i(x''k'')$$

where x' and x'' signify two space-time points, k' and k'' two spin indices. The sum should extend over all occupied states. From this matrix, one can then easily form the current- and charge-densities $\vec{i}$, ρ and the energy–momentum tensor[17] U^μ_ν, on the basis of the relations

$$\vec{i} = \lim_{x'=x''} \sum_{k'k''} (\vec{\alpha})_{k'k''}(x'k'|R|x''k'')$$

$$\rho = \lim_{x'=x''} e\sum_{k'} (x'k'|R|x''k'')$$

$$U^\mu_\nu = \lim_{x'=x''} \frac{1}{2}\left\{ich\left[\frac{\partial}{\partial x'_\mu} - \frac{\partial}{\partial x''_\mu}\right] - e[A^\mu(x') + A^\mu(x'')]\right\}$$

$$\sum_{k'k''} (\alpha^\nu)_{k'k''}(x'k'|R|x''k'').$$

$$\alpha^4 = \text{unit matrix}.$$

The density matrix has the advantage that for $x' \neq x''$, the summations over the vacuum electrons do not diverge, but yield an expression that becomes singular for $x' = x''$.

From the assumption I, one can now clearly specify which parts of the density matrix of the vacuum electrons go over into omitted parts for $x' = x''$, and in this fashion one obtains an explicit representation of these terms.

The physically meaningless portion of the density matrix must then consist of those terms which are independent of the field strengths, of those which lead to terms in the current density proportional to the derivatives of the fields, and of those which lead to

[15] H. Bethe and W. Heitler, *Proc. Roy. Soc.* **146**, 84, 1934.
[16] H. R. Hulme and J. C. Jaeger, *Proc. Roy. Soc.* **153**, 443, 1936.
[17] The complete energy–momentum tensor consists of the sum of U^μ_ν and the Maxwell energy–momentum tensor of the field. The U^4_4-component is thus not the total material energy density, but only the kinetic one.

terms in the energy density which are proportional to the square of the field strength. Furthermore, the part of the density matrix that ought to be subtracted must also be multiplied by the factor

$$u' = \exp\left[\frac{ie}{hc}\int_{x'}^{x''}\left(\sum_{i=1}^{3} A_i \, dx_i - V \, dt\right)\right],$$

where the integral in the exponent extends in a straight line from the point x' to the point x''. This factor adds to the energy–momentum tensor being subtracted precisely from those contributions which come from the fact that the still unperturbed vacuum electrons in the field obtain an additional energy eV and an additional momentum $(e/c)\vec{A}$, which, due to assumption I_2, must also be subtracted.

Since the portion of the density matrix that must be subtracted is at most of second order in the field strengths, except for the factor u', it can be obtained from the density matrix of the free electrons by means of a perturbation calculation. This calculation is simple in principle, but very complicated in execution, and it forms the basis for the determination of this matrix by Heisenberg *loc. cit.* The result can be formulated mathematically more simply if one forms the average for every quantity from the calculation by means of the present theory and from the calculation by means of a theory in which the electron charge is positive and the negative electron is represented as a 'hole'. The result is really the same in both cases. The density matrix R is then replaced by R':

$$(x'k'|R'|x''k'') = \frac{1}{2}\left\{\sum_i \psi_i^*(x'k')\psi_i(x''k'') - \sum_k \psi_k^*(x'k')\psi_k(x''k'')\right\},$$

where the first sum extends over the occupied states and the second sum over the unoccupied states.

The portion $(x'k'|S|x''k'')$ which is to be subtracted, then has the form

$$(x'k'|S|x''k'') = u'S_0 + \frac{\bar{a}}{|x' - x''|^2} + \bar{b}\lg\frac{|x' - x''|^2}{C}.$$

Here, S_0 is the matrix R' for vanishing potentials, $\bar{a}$ and $\bar{b}$ are functions of the field strengths and their derivatives, C is a constant. These quantities are explicitly given in Heisenberg, *loc. cit.* and in Heisenberg and Euler, *loc. cit.*

To perform special calculations, it is more practical not to fall back on Heisenberg's explicit expression, but to recognize the deleted terms by their structure. This is simpler especially because the remaining expressions no longer become singular at $x' = x''$, so that one really does not need the formal aid of the density matrix to calculate them. Summation over all vacuum electrons here no longer leads to divergent expressions. However, the explicit representation of Heisenberg is well suited to exhibit the relativistic invariance and the validity of the conservation laws in this method.

From this it can be seen that the determination of the physical properties of the vacuum electrons, such as has been done here, essentially does not contain any arbitrariness, since only and exclusively those effects of the vacuum electrons are omitted which must be omitted due to the basic assumption of positron theory; the energy and the charge of vacuum electrons, unperturbed by the field, and the physically meaningless field-independent constant polarizability of the vacuum. All physically meaningful effects of the vacuum electrons are taken into account and lead to convergent expressions. From this, one may surely draw the conclusion that the hole theory of the positron has brought with it no essential difficulties for electron theory as long as one limits oneself to treating unquantized wave fields.

At this point, I would like to express my sincerest thanks to Professors Bohr, Heisenberg, and Rosenfeld for many discussions. I am also grateful to the Rask-Ørsted Fund, which has made it possible for me to perform this research at the Institute for teoretisk Fysik in Copenhagen.

9 Theory of the emission of long-wave light quanta

W. PAULI and M. FIERZ

Nuovo Cimento, 15, 167–88. Presented at the Galvani Bicentenary Congress, Bologna, 18–21 October 1937.

Summary – As is well known, the usual radiation theory yields an infinitely large value (infrared catastrophe) for the cross section $\mathrm{d}q$ of a charged particle when it traverses a force field and is deflected at a given angle. For if one prescribes that the energy loss of the particle should lie between E and $E + \mathrm{d}E$, then, according to this theory, for small E one obtains $\mathrm{d}q =$ const. $\mathrm{d}E/E$, which, upon integration, diverges logarithmically at the point $E = 0$. The present paper investigates more precisely what quantum electrodynamics yields for this cross section if a finite extent is ascribed to the charged body. It appears that then the infinity is indeed removed and that the deflections which are considered as nonradiative in the ordinary theory here appear as having a finite, although very small, energy loss. On the other hand, according to the exact theory, the more precise behavior of $\mathrm{d}q$ for very small energy losses E depends so strongly on the extension of the charged body, that a direct application of the result to real electrons is not possible. Therefore, one must conclude that the problem in question is related in an essential manner to the still unresolved fundamental difficulties of quantum electrodynamics.

§1 Introduction

One of the well-known difficulties of quantum electrodynamics concerns the infinite self-energy of a charged particle. In addition, as is well known, there is also a divergent result of this theory which concerns the emission of light quanta of very low frequency. According to this theory, the probability of finding a charged particle deflected in a certain direction when it passes through a force field, where the particle simultaneously suffers an energy loss lying in the interval $(E, E + \mathrm{d}E)$, turns out to be proportional to $\mathrm{d}E/E$ for small E. Integration of this probability over a finite interval (E, E') consequently yields a logarithmic infinity at $E = 0$. The present paper shall investigate whether this result is caused only by unallowable mathematical approximations or whether a more profound physical difficulty is involved. To answer this question, we have used quantum electrodynamics for rigid, spatially extended charged bodies. As is well known, this always leads to finite results for a nonvanishing extension a of the body – which is equivalent (up to a numerical factor of order 1) to a 'cutoff frequency' that is given by $\omega_1 = 2\pi c/a$. We need introduce no assumption about the value of

e^2/hc, where e denotes the charge of the particle. However, to avoid complicating the problem by the appearance of the electromagnetic mass $\mu \sim e^2/ac^2 \sim e^2\omega_1/c^3$, and simultaneously to make the definition of the mass of the body unambiguous, we first of all introduce the explicit presupposition that μ is assumed to be small compared to the mechanical mass m of the body, or

$$\frac{e^2\omega_1}{mc^3} \ll 1. \tag{I}$$

The mere fact that we must introduce a finite dimension for the charged body and, as we shall see, that the final result depends on this dimension, made it seem logical to us (because of the Lorentz contraction) to assume in advance that the initial speed v_0 of the body is small compared to the speed of light:

$$\frac{v_0}{c} \ll 1. \tag{II}$$

We shall always calculate with nonrelativistic wave mechanics. Further, it is suitable also to introduce the presupposition

$$\frac{h\omega_1}{mc^2} \ll 1, \tag{III}$$

which is more stringent than (I) for small e^2/hc. As a consequence, the Compton shift is relatively small for all proper vibrations of the radiation cavity, which interact with the body. That is, the momentum of the radiation (for arbitrary values of e^2/hc) can be neglected compared to the change of the particle momentum. It is thus possible to treat the radiation as dipole radiation, i.e. in the Hamiltonian operator to replace the vector potential of the radiation field at the position of the particle by that at a fixed position.

A strict solution of the reduced problem defined in this way, in the case where external fields are absent, as well as an approximate solution in the case of weak external fields, has already been given by Bloch and Nordsieck[1], who succeeded in making significant progress in this manner. Section 2 of the present paper contains nothing new compared with the work of these authors and has been presented here in detail for the sake of continuity, so as to so show clearly the assumption of the finite dimension of the charged body as well as the simplifications, which are made possible by presupposition (II), as compared to the somewhat more general assumptions of the above authors.[2]

The essential result is that in the approximation considered here, which is characterized by presuppositions (I) to (III), the radiation field can be characterized by 'free' light quanta even when the charged particle is present. In the absence of external forces, the number of these light quanta remains constant. Now, if the body traverses an external force field, completely nonradiative deflections of the particle never occur; rather, an infinite number of free light quanta are always emitted. However, nearly all of these have very low frequencies, since the total emitted radiative energy is finite.

Unfortunately, we did not succeed in solving the mathematical problem defined in §2 for arbitrary force fields; in particular, a more precise investigation of the limiting process to classical mechanics and electrodynamics would still be desirable. However, in the present paper we limit ourselves to a discussion of the special case of weak force

[1] F. Bloch and A. Nordsieck, *Phys. Rev.*, **52**, 54, 1937, quoted below as 'A'. A. Nordsieck, *Phys. Rev.*, **52**, 59, 1937, quoted below as 'B'.

[2] Compare 'Remark' [A], p. 8. Compare note 6, p. 232.

fields, where one can limit oneself to the first approximation of perturbation theory (Born approximation). In contrast to Bloch and Nordsieck, however, we deem it essential, in the discussion of the theoretical results, to take conservation of energy into account *exactly*. A comprehensive discussion of the results of quantum electrodynamics for the cross section $\mathrm{d}q$ of the particle, for deflection in a given angular region $\mathrm{d}\Omega$ and a given energy loss of the particle, as well as for the emitted spectrum, is given in §3 and §4 of the present paper.

Here, we only wish to give a special result for the case

$$\frac{e^2}{hc}\cdot\frac{v_0^2}{c^2} \ll 1 \quad \text{and} \quad \frac{m}{2}v_0^2 \ll h\omega_1,$$

which is important for the basic question that was raised initially. If $\vec{v}_0$ and $\vec{v}_0'$ are the initial and final velocities of the particle, and if we use the abbreviation

$$C = \frac{2}{3\pi}\frac{e^2}{hc}\frac{(\vec{v}_0 - \vec{v}_0')^2}{c^2}$$

the above cross section becomes

$$\mathrm{d}q = \mathrm{const}\cdot\frac{\mathrm{d}E}{E}\cdot\left(\frac{E}{h\omega_1}\right)^C,$$

where the constant factor for a given deflection angle is the same as in previous radiation theory. As a consequence, this expression now becomes integrable at $E = 0$; in particular, the total probability of finding the particle with an energy loss in the finite interval $(0, E)$ and deflected in a given direction turns out to have a magnitude which agrees very well with the result of the old theory for nonradiative transitions, as long as E is small compared to the initial energy $E_0 - (m/2)v_0^2$ of the particle but is large compared to the critical energy

$$\varepsilon = h\omega_1\cdot\exp\left(-\frac{1}{C}\right).$$

However, if the energy loss of the particle is of the same order of magnitude as the quantity ε (which is small by presupposition), the cross section depends significantly on ω_1, i.e. on the size of the charged particle, although the wavelength of the light is really very large compared to the dimensions of the body. With the present state of quantum theory, it cannot be decided with certainty to what extent this result of the quantum electrodynamics for extended bodies corresponds to reality, since this result exceeds the framework of ordinary correspondence considerations. In any case, an improvement of the ordinary radiation theory (in accordance with the correspondence principle) for point charges or for real electrons is probably not possible, even for the problem considered here, without going into the fundamental self-energy difficulties of quantum electrodynamics.

§2 The reduced radiation problem

The Hamiltonian function of the nonrelativistic theory at first reads as follows:

$$H = \frac{1}{2m}\left(\vec{p} - \frac{e}{c}\vec{A}\right)^2 + V(x) + \frac{1}{8\pi}\int(\vec{E}_{\mathrm{tr}}^2 + \vec{H}^2)\,\mathrm{d}v. \tag{1}$$

Here E_{tr} is the divergence-free part of the electric field strength, $V(x)$ is the potential of the external force field, and e and m are the charge and mechanical mass of the charged body. If the latter has a finite extension, one must here take for $\vec{A}$ the average of the potential over the charge distribution of the body, which is thought of as being rigid:

$$\vec{A} = \int \vec{A}(\vec{x} + \vec{\xi}) D(\vec{\xi}) \, \mathrm{d}^3\xi \tag{2}$$

with the normalization

$$\int D(\xi) \, \mathrm{d}^3\xi = 1. \tag{3}$$

For the sake of simplicity, we assume spherical symmetry for the body. As a consequence, D depends only on the magnitude $\rho = |\vec{\xi}|$ of the vector $\vec{\xi}$. If one decomposes $\vec{A}$ into proper vibrations in a finite cavity with the volume V and with cyclic boundary conditions, one can set

$$\frac{e}{mc} \vec{A} = \sum_s h\omega_s \vec{a}_s [P_s \cos(\vec{k}, \vec{x}) + Q_s \sin(\vec{k}, \vec{x})] \tag{4a}$$

[see note 3, below]

$$\frac{1}{8\pi} \int (\vec{E}_{\text{tr}}^2 + \vec{H}^2) \, \mathrm{d}V = \frac{1}{2} \sum_s (P_s^2 + Q_s^2) h\omega_s. \tag{4b}$$

Here,

$$\vec{a}_s = \frac{2e}{m} \cdot \frac{1}{h\omega_s} \left(\frac{\pi h}{\omega_s \cdot V} \right)^{1/2} \vec{\varepsilon}_s g(\vec{k}_s) \tag{5}$$

[see note 4, below]. $\vec{\varepsilon}_s$ is a polarization unit vector perpendicular to $\vec{k}_s$, while the P_s and Q_s are operators obeying the commutation relations

$$[P_s, Q_{s'}] = -\mathrm{i}\delta_{ss'}; \, [P_s, P_{s'}] = [Q_s, Q_{s'}] = 0$$

and which also commute with $\vec{x}$ and $\vec{p}$.

The factor $g(\vec{k}_s)$ in (5) is the 'cutoff factor' that is based on the finite shape of the body, and which is related to the previously introduced function $D(\xi)$ according to

$$g(\vec{k}_s) = \int D(\xi) \exp(\mathrm{i}\vec{k}_s \cdot \vec{\xi}) \, \mathrm{d}^3\xi;$$

$$D(\vec{\xi}) = \frac{1}{(2\pi)^2} \int g(\vec{k}_s) \exp(-\mathrm{i}\vec{k}_s \cdot \vec{\xi}) \tag{6}$$

According to our presupposition of spherical symmetry for the body, it depends only on the magnitude $|k_s| = \omega_s/c$. The normalization condition (3) directly yields

$$g(0) = 1, \tag{3a}$$

while

$$\int_0^\infty G(\omega) \, \mathrm{d}\omega \equiv \int_0^\infty |g(\omega)|^2 \, \mathrm{d}\omega = \omega_1 \tag{7}$$

[3] By h we always designate Planck's constant divided by 2π, and ω designates the circular frequency.

[4] The quantity which is here designated by $\vec{a}_s$ differs from that in the Bloch and Nordsieck paper [A] *loc. cit.* by the factor $1/mh\omega_s$.

shall be designated for short as the 'cutoff frequency'. c/ω_1 is proportional to the dimension of the body, so that $\omega_1 = \infty$ corresponds to a point-shaped body. Functions $g(\omega)$, which are especially convenient for the calculations, are

$$g(\omega) = 1 \text{ for } 0 < \omega < \omega_1;\ g(\omega) = 0 \text{ for } \omega > \omega_1$$

and

$$|g(\omega)|^2 \equiv G(\omega) = \mathrm{e}^{-\omega/\omega_1}. \tag{8}$$

Corresponding to our presuppositions (I) and (III), we can now first of all neglect the term proportional to A^2 in the Hamiltonian operator and, secondly, we can replace the center point x of the body in the argument of the vector potential by the fixed center point of the force field $V(x)$, which at the same time we let coincide with the origin of the coordinate system. This means that in (4) we set $(\vec{k}, \vec{x}) = 0$, and thus obtain for the simplified Hamiltonian operator

$$H = \frac{p^2}{2m} - \sum_s h\omega_s(\vec{a}_s\vec{p})P_s + V(x) + \sum_s \tfrac{1}{2}(P_s^2 + Q_s^2)h\omega_s. \tag{9}$$

This means that we treat radiation as electric dipole radiation and neglect the higher poles.

Now, in the force-free case $V(x) = 0$, we can solve the problem by a simple canonical transformation, namely

$$\begin{cases} \vec{x} = \vec{x} - h\sum_s \vec{a}_s Q_s & \vec{p} = \vec{p}' \\ Q_s = Q'_s & P_s = P'_s + (\vec{a}_s\vec{p}). \end{cases} \tag{10}$$

Then (9) becomes

$$H = \frac{p^2}{2m} - \frac{1}{2}\sum_s (\vec{a}_s\vec{p})^2 h\omega_s + V(\vec{x}' - h\sum_s \vec{a}_s Q_s)$$
$$+ \sum_s \sum_1^2 (P_s'^2 + Q_s^2)h\omega_s.$$

It can readily be seen that the second term corresponds to the presence of the electromagnetic mass. In fact, performing the summation over s according to (5), and taking into account the fact that the number of polarized proper vibrations between ω_s and $\omega_s + \mathrm{d}\omega_s$ has the value

$$Z_s = \frac{V}{\pi^2 c^3} s_s^2 \,\mathrm{d}\omega_s \tag{11}$$

yields

$$\sum_s (\vec{a}_s\vec{p})^2 h\omega_s = \vec{p}^2 \frac{4}{3\pi} \cdot \frac{e^2}{m^2c^3} \int_0^\infty \mathrm{d}\omega G(\omega) = p^2 \frac{4}{3\pi} \frac{e^2\omega_1}{m^2c^3}.$$

Upon introducing the electromagnetic mass[5]

$$\mu = \frac{4}{3} \frac{e^2\omega_1}{\pi e^3} \tag{12}$$

[5] On the basis of (6) and (7), one easily finds for the electrostatic energy of the charge distribution $eD(\xi)$ the value $w = e^2\omega_1/\pi c$, so that (12) reproduces the familiar relation $\mu = \frac{4}{3}w/c^2$ for spherically symmetric charge distributions.

one thus obtains

$$\frac{p^2}{2m} - \frac{1}{2}\sum_s (a_s p)^2 h\omega_s = \frac{p^2}{2m}\left(1 - \frac{\mu}{m}\right).$$

According to our presupposition (I) it is logical to neglect the second term here, and we now obtain the Hamiltonian operator of the finally reduced problem, as follows:

$$H = \frac{p^2}{2m} + V(\vec{x}' - h\sum_s \vec{a}_s Q_s) + \sum_s \tfrac{1}{2}(P_s'^2 + Q_s^2)h\omega_s. \qquad \text{(IV)}$$

In the force-free case $V(x) = 0$, we obtain the solutions

$$\psi_0(x', Q_s) = \exp\left(\frac{i}{h}\vec{p}\cdot\vec{x}\right)\prod_s h_{n_s}(Q_s), \qquad (13)$$

where the $h_n(Q)$ are the normalized orthogonal functions which are associated with the hermite polynomials and which satisfy the equation

$$-\frac{d^2 h_n}{dQ^2} + Q^2 h_n = (2n+1)h_n.$$

The quantum numbers n_s, which are defined by

$$\tfrac{1}{2}(P_s'^2 + Q_s^2) = n_s + \tfrac{1}{2} \qquad (14)$$

can be designated as the numbers of the 'free light quanta' of frequency ω_s, and differ from the numbers of light quanta

$$\tfrac{1}{2}(P_s^2 + Q_s^2) = N_s + \tfrac{1}{2},$$

which are used elsewhere in radiation theory. But from the Hamiltonian operator it follows that – in the approximation used here – the numbers n_s of free light quanta are the ones which can be directly observed with a spectrograph, since only these can propagate freely at a spatial separation from the charged body. In particular, for a given particle momentum $\vec{p}$, the state of lowest energy, in which all quantum numbers n_s vanish, corresponds to the absence of radiation in the true sense. The normalization of the eigenfunctions (13) relative to coordinate space is such that the average density of the particle in large spaces becomes equal to 1, which is suitable for the collision problems being treated here. As can easily be seen, these functions form a complete orthogonal system with respect to the parameters p and n_s.[6]

In general, therefore, we can introduce $\vec{p}$ and the n_s as arguments of the wave function in accordance with the formulation

$$\psi(x', Q) = \sum_{n_s}\int d^3p\,\varphi(p, n_s)\exp\left(\frac{i}{h}(\vec{p}, \vec{x})\right)\prod_s h_{n_s}(Q_s). \qquad (15)$$

[6] In A it was attempted to treat the force-free case more strictly, without identifying the particle position x with a fixed location. The resultant functional system, Equation (17) in A, which appears as a generalization of our system (13), is not orthogonal, however, as appears directly from a specialization of Equations (21)–(23) of A for $V(x) = 1$. However, there is approximate orthogonality for the matrix elements between such states for which the total momentum $|\sum n_s \vec{k}_s|$ of the radiation is very small compared to the momentum change $|\vec{p} - \vec{p}'|$ of the particle. In cases where only matrix elements of $V(x)$ between such states significantly influence the final results, the more general formulas in A consequently should represent a useful approximation. Under our further presupposition (II), and especially (III), on the other hand, the method pursued here seems more logical, where we consider only electric dipole radiation for light emission.

According to (IV), in the presence of forces (omitting the zero-point energy of the radiation), and in a stationary state of the energy E, the wave functions $\varphi(\vec{p}, n_s)$ satisfy the Schrödinger equation

$$E\varphi(\vec{p}, n_s) = \left(\frac{p^2}{2m} + h\sum_s n_s\omega_s\right)\varphi(p, n_s)$$

$$+ \sum_{n'_s}\int d^3p'(p|V|p')\prod_s(n_s, p|K|n'_s, p')\varphi(p', n'_s) \tag{16}$$

with

$$(p|V|p') = \frac{1}{(2\pi h)^2}\int \exp\left(\frac{i}{h}(\vec{p}' - \vec{p})\cdot\vec{x}\right)V(x)\,d^3x \tag{17}$$

$$(n_s, p|K|n'_s, p') = \int \exp(i(\vec{p}' - \vec{p})\cdot\vec{a}_s)Q_s h_{n'_s}(Q)h_{n_s}(Q)\,dQ. \tag{18}$$

Using the abbreviation

$$w_s = \frac{1}{2}[(\vec{p}' - \vec{p})\cdot\vec{a}_s]^2 \tag{19}$$

one obtains

$$(n, p|K|n', p') = \exp\left(-\frac{w}{2}\right)(n!n'!)^{1/2}\sum_{k=0}^{\infty}\frac{(iw^{1/2})^{n'-n+2k}}{(n-k)!k!(n'-n+k)!}. \tag{20}$$

If one sets $1/(n' - n + k)! = 0$ for $n' - n + k < 0$, this equation holds both for $n' \geqq n$ and for $n' \leqq n$, and furthermore the symmetry of the expression in n and n' becomes apparent through the substitution $k' = n' - n + k$. For $\vec{p}' = \vec{p}$, K differs from zero only if $n = n'$, and in this case it is equal to 1. Furthermore, there particularly holds

$$(n, p'|K|0, p') = \exp\left(-\frac{w}{2}\right)(iw^{1/2})^n\frac{1}{(n!)^{1/2}}. \tag{20a}$$

Upon summing over all n', there immediately follows from (18), through the completeness relation for the $h_n(Q)$,

$$\sum_{n'_s}(n, p|K|n', p')(n', p'|K|n'', p'') = (n, p|K|n'', p'') \tag{21}$$

and especially for $p'' = p$, $n'' = n$:

$$\sum_{n'}|(n, p|K|n', p')|^2 = 1. \tag{21a}$$

For $n' = 0$, this also follows easily directly from (20a).

We shall here be interested in a special eigenvalue problem associated with (16), which corresponds to a plane wave without a light quantum, which is incident in the $+x_1$ direction and which is scattered by the force field. Consequently, here in particular there holds

$$\psi(x', Q) = \exp\left(\frac{i}{h}p_0x'_1\right)\prod_s h_0(Q_s) + \psi_1(x', Q_s), \tag{22}$$

where $\psi_1(x', Q_s)$ may contain only *outward propagating* spherical waves for all Q_s and for large $\vec{x}'$ (or $\vec{x}$) (radiation condition). Corresponding to the decomposition (22), we obtain from (15) and (16)

$$\varphi(\vec{p}, n_s) = \delta(\vec{p} - \vec{p}_0)\prod_s \delta(n_s, 0) + \frac{f(\vec{p}, n_s)}{\frac{p_0^2 - p^2}{2m} - h\sum_s n_s\omega_s} \tag{23}$$

where we have used

$$E = p_0^2/2m.$$

Here, f must fulfil the equation

$$f(\vec{p}, n_s) = (p|V|p_0)\prod_s (n_s, p|K|0, p_0)$$

$$+ \sum_{n'_s}\int d^3p'(p|V|p')\frac{\prod_s (n_s, p|K|n'_s, p')f(n'_s, p')}{\frac{p_0^2 - p'^2}{2m} - h\sum n'_s\omega_s}. \tag{24}$$

In familiar fashion, the radiation condition is automatically satisfied as follows. In the integration over p or p' space, one deforms the integration path as follows *into the complex plane*: One should integrate in polar coordinates, perform the angle integration in the standard way, but when integrating over the absolute magnitude $|\vec{p}| = P$ or $|\vec{p}'| = P'$, one detours around the poles of the integrand which lie on the positive real axis by going into the *lower* complex P or P' half plane. From (23) one can easily calculate[7] that, for large $|\vec{x}'| = r$, a scattered wave

$$A \cdot \frac{\exp\left(\frac{i}{h} p'_0 r\right)}{r}$$

results, whose momentum is given by the energy condition[8]

$$\frac{1}{2m} p_0'^2 = \frac{1}{2m} p_0^2 - h\sum n_s\omega_s \tag{25}$$

and whose amplitude is

$$A = 4\pi^2 hmf(|p'_0| \cdot \vec{e}, n_s), \tag{26}$$

where $\vec{e} - \vec{x}/r$ is a unit vector in the emission direction under consideration. For given quantum numbers n_s of the emitted free light quanta, therefore, the cross section dq for scattering the particle in the direction $\vec{e}$ per angular range $d\Omega$ is given by

$$dq = d\Omega \cdot 16\pi^4 h^2 m^2 \frac{|p'_0|}{|p_0|} |f(|p'_0| \cdot \vec{e}, n_s)|^2. \tag{27}$$

[7] In this connection also compare P. A. M. Dirac, *Quantum Mechanics*, 2nd Edition, §53, particularly p. 200, Equations (37), (38). The quantities here designated by f differ from the Dirac quantities by the factor $(2\pi h)^{-3/2}$.

[8] For values of n_s, for which the right side of (25) is negative, $\psi(x')$ for large r drops off faster than $1/r$.

§3 Scattering of the body and light emission in weak force fields

Major mathematical difficulties prevent a general discussion of the problems defined by Equations (23) and (24) (and the associated integration prescription). Consequently, we shall here limit ourselves to a force field that is so weak that it can be regarded as a small perturbation on the force-free problem.[9] Accordingly, we shall here retain in the perturbed wave function only terms of first order in $V(x)$ (Born approximation). But it should be noted that nothing stands in the way of extrapolating this method and that the 'collision time' defined in classical mechanics is quite irrelevant as a justification to stop the perturbation calculation at the first approximation.[10] The first approximation of perturbation theory, upon development in terms of the force potential, follows directly from (24):

$$f(\vec{p}, n_s) = (p|V|p_0) \cdot \prod_s (p, n_s|K|p_0, 0)$$

since the next term is already of second order. Inserting (19) and (20a), and taking (25) into account, yields

$$|f(\vec{p}_0', n_s)|^2 = (p_0'|V|p_0)|^2 \prod_s \frac{1}{n_s!} w_s^{n_s} \exp - w_s. \tag{28}$$

The form of the dependence of the right side of n_s means that the emission of various quanta (of the same or different frequency) for given p_0 and p_0' is *statistically* independent. This appears to us to be characteristic of weak force fields.

The fact that the total cross section for a given direction of deflection of the particle diverges in the usual radiation theory is directly connected with the fact that the infinite product in (28) is zero as long as only a finite number of the n_s are different from zero; according to (5) and (11), the sum $\sum_s w_s^2$ diverges for small ω_s like $\int d\omega_s/\omega_s$. In A, this led to the conclusion that infinitely small quanta $h\omega_s$ are emitted during each passage of the charged body through the force field and that, in particular, there could be no completely elastic collisions where the energy loss of the particle vanishes exactly.

Beyond this, we shall first calculate the exact expression for the cross section dq, however, if – with any light emission whatsoever – the energy loss

[9] This is in harmony with the method pursued in A. In B, an attempt is made to attack the problem for general force fields – at least under the restrictive assumption $e^2/\hbar c \ll 1$ – by starting from the wave functions $\varphi(p, n_s) = \varphi^0(p)\prod_s (n_s, p|k|0, p_0)$ as a first approximation. Here $\varphi^0(p)$ is the exact solution of the non-radiative problem. That is, $\varphi^0(p)$ satisfies the equations $\varphi_0(p) = \delta(\vec{p} - \vec{p}_0) + f^0(\vec{p})/(1/2m)(p_0^2 - p^2)$ with $f^0 = (p|V|p_0) + \int d^3p(p|V|p')f_0(p')/(1/2m)(p_0^2 - p'^2)$. However, since this wave function $\varphi(p, n_s)$ contains only elastically reflected spherical waves (compare the difference of the energy denominators in $\varphi^0(p)$ and in (23)), it is not allowable, starting from such a wave function, to use an ordinary perturbation method. Consequently, up to now it has not been demonstrated that the method used in B – apart from the case of weak force fields, where it coincides with the method of A – leads to correct results.

[10] For the special force field $V(r) = (h^2/2m) \cdot K \exp(-\alpha r)/r$ (where K has the dimension of a reciprocal length), we have determined the second approximation of perturbation theory and have observed that (for arbitrary values of the charge of the body) it remains small compared to the first one, as long as $K \gg \alpha$. However, the Born approximation fails for the Coulomb field ($\alpha = 0$) as soon as $|p_0'| \sim hK$, that is particularly in the neighborhood of the edge of the brems-spectrum ($p_0' = 0$), as is indeed well known.

$$\frac{1}{2m}(p_0^2 - p_0'^2) = h\sum_s n_s\omega_s$$

of the body, for a given deflection angle, lies between E and $E + \mathrm{d}E$:

$$E < h\sum_s n_s\omega_s < E + \mathrm{d}E. \tag{29}$$

According to (27) and (28), this is determined by

$$\mathrm{d}q = \mathrm{d}\Omega \cdot 16\pi^4 h^2 m^2 \frac{|p_0'|}{|p_0|} |(p_0'|V|p_0)|^2 \cdot S(E)\,\mathrm{d}E \tag{30}$$

with

$$S(E)\,\mathrm{d}E = \sum_{E<h\Sigma n_s\omega_s<E+\mathrm{d}E} (n_s)\prod_s \frac{1}{n_s!} w_s^{n_s} \cdot \exp(-w_s). \tag{31}$$

This expression can be evaluated in the following fashion. By introducing the Dirac δ-function, one can first of all write

$$S = \sum(n_s)\delta(E - h\sum_s n_s\omega_s) \cdot \prod_s \frac{1}{n_s!} w_s^{n_s} \exp(-w_s),$$

where now the sum is to be extended over *all* n_s.[11] Then one has

$$\delta(x) = \frac{1}{2\pi}\int_{-\infty}^{+\infty} \exp(\mathrm{i}\alpha x)\,\mathrm{d}\alpha.$$

If we insert this and interchange the integration over α with the summation over the n_s, there follows:

$$S = \frac{1}{2\pi}\int_{-\infty}^{+\infty} \mathrm{d}\alpha \exp(\mathrm{i}\alpha E)\prod_s \sum_{n_s=0}^{\infty} \frac{1}{n_s!} w_s^{n_s} \exp(\mathrm{i}\alpha h\omega_s n_s)\exp(-w_s)$$

or, performing the summation over n_s,

$$S = \frac{1}{2\pi}\int \mathrm{d}\alpha \exp(\mathrm{i}\alpha E + f(\alpha)) \tag{32}$$

with

$$f(\alpha) = \sum_s w_s[\exp(-\mathrm{i}\alpha h\omega_s) - 1].$$

By means of (11), and after inserting expression (5), the sum can be converted into an integral, and one obtains

$$f(\alpha) = C \cdot \int_0^\infty \frac{\mathrm{d}\omega}{\omega}[\exp(-\mathrm{i}\alpha h\omega)] \cdot G(\omega) \tag{33}$$

wherein we insert

$$C = \frac{2}{3\pi}\frac{e^2}{hc} \cdot \frac{(\vec{p}_0 - \vec{p}_0')^2}{m^2c^2} \tag{34}$$

[11] In A, $h\sum n_s\omega_s$ is here neglected in the argument of the δ-function, which does not appear logical to us.

and $G(\omega)$ again signifies the cutoff factor. One sees first of all that, without the cutoff factor ($G(\omega) = 1$, point charge) $f(\alpha) = -\infty$ and $S = 0$. *The value of $S(E)$ consequently must depend essentially on ω_1 and generally also on the form of the function $G(\omega)$.* Later, special cases will be discussed for general $G(\omega)$. First we shall perform the calculation for the choice $G(\omega) = \exp(-\omega/\omega_1)$, which has been given in (8). In this case, $f(\alpha)$ can readily be evaluated in closed form as follows:[12]

$$f(\alpha) = -C\lg(1 + \mathrm{i}\alpha h\omega_1) = \lg(1 + \mathrm{i}\alpha h\omega_1)^{-C} \text{ for } G(\omega) = \exp(-\omega/\omega_1). \tag{35}$$

In this case, (32) becomes

$$S = \frac{1}{2\pi}\int_{-\infty}^{+\infty} \exp(\mathrm{i}\alpha E)(1 + \mathrm{i}\alpha h\omega_1)^{-C}\,\mathrm{d}\alpha$$

and with

$$1 + \mathrm{i}\alpha h\omega_1 = \frac{h\omega_1}{E}\cdot z;\ \exp(\mathrm{i}\alpha E) = \exp z\exp(-E/h\omega_1);\ \mathrm{d}\alpha = \frac{1}{E}\frac{\mathrm{d}z}{\mathrm{i}}$$

$$S = \exp(-E/h\omega_1)\frac{1}{h\omega_1}\left(\frac{h\omega_1}{E}\right)^{C-1}\cdot\frac{1}{2\pi\mathrm{i}}\int z^{-C}\exp z\,\mathrm{d}z.$$

The integration path in the integral over $\mathrm{d}z$ can be deformed so that, leaving the origin at the left, it runs from $-\infty - \mathrm{i}\varepsilon$ to $-\infty + \mathrm{i}\varepsilon$, and according to a well-known formula one obtains

$$\frac{1}{2\pi\mathrm{i}}\int z^{-C}\exp z\,\mathrm{d}z = \frac{1}{\Gamma(C)}$$

$$S = \frac{1}{h\omega_1}\left(\frac{E}{h\omega_1}\right)^{C-1}\frac{\exp(-E/h\omega_1)}{\Gamma(C)} \text{ for } G(\omega) = \exp(-\omega/\omega_1). \tag{36}$$

Another interesting physical quantity can be reduced to this same function $S(E)$. We call this quantity the spectral *emission coefficient*. This specifies the energy in the frequency interval $(\omega, \omega + \mathrm{d}\omega)$ which is emitted under stationary conditions by a single incident particle that passes through a unit surface (perpendicular to the x_1 axis). It should be designated by $A(\omega)\,\mathrm{d}\omega$.[13] Here we can also specify the direction of the particle deflected during the process (as lying in a particular solid angle of magnitude $\mathrm{d}\Omega$). We can also specify first the energy loss of the particle (as lying in the interval E, $E + \mathrm{d}E$); subsequently we shall integrate over the latter. We here consider the energy emitted in this frequency interval $(\omega, \omega + \mathrm{d}\omega)$, *regardless of what other frequencies are also emitted at the same time during the processes under consideration. Furthermore, we leave the direction of the emitted quanta arbitrary*. From (27) and (28) there now directly follows, for given $\vec{p}_0'$,

[12] Note that $\int_0^\infty \mathrm{d}x/x(\exp(-x) - \exp(-ax)) = \lg a$.

[13] If J_0 is the current density (the number of particles passing through the unit surface per unit time) of the incident particles, $J_0A(\omega)\,\mathrm{d}\omega$ should be the light energy emitted per unit time in the processes under consideration. If N is the number of scattering centers per unit volume, $-\mathrm{d}E/\mathrm{d}x = E\cdot N\int_0^\infty A(\omega)\,\mathrm{d}\omega$ is the total average energy loss of the incident particles of energy E per unit length. $A(\omega)\,\mathrm{d}\omega$ has the dimension erg cm^2.

$$A(\omega)\,\mathrm{d}\omega = \mathrm{d}\Omega\cdot 16\pi^4 h^2 m^2 \frac{|p_0'|}{|p_0|}|(p_0'|V|p_0)|^2$$

$$\cdot \prod_{\substack{0<\omega_s<\omega \text{ and}\\ \omega_s>\omega+\mathrm{d}\omega}} (s)\sum(n_s)\frac{1}{n_s!}w_s^{n_s}\exp - w_s$$

$$\cdot \prod_{\omega<\omega_s<\omega+\mathrm{d}\omega} (s)\sum(n_s) n_s h\omega_s \frac{w_s}{n_s!}\exp - w_s.$$

Here, the n_s must always satisfy the condition

$$E < h\sum n_s\omega_s < E + \mathrm{d}E,$$

which expresses conservation of energy, and which must be taken into account as above by introducing the δ-function. The direct consequence of this condition is that necessarily

$$h\omega \leqslant E \tag{37}$$

so that the radiative energy emitted at frequency ω can be different from zero. By means of the same method as above, one obtains directly

$$A(\omega)\,\mathrm{d}\omega = \mathrm{d}\Omega\cdot 16\pi^4 h^2 m^2 \frac{|p_0'|}{|p_0|}|(p_0'|V|p_0)|^2\cdot S(E - h\omega)\,\mathrm{d}E$$
$$\cdot C\cdot G(\omega)\cdot h\,\mathrm{d}\omega. \tag{38}$$

The value of the function S for the argument $E - h\omega$ thus enters here, C is again given by (34), and one must always insert

$$|p_0'| = (p_0^2 - 2mE)^{1/2}. \tag{39}$$

If one specifies only the direction of the particle momentum after emission, one must integrate with this value of $|p_0'|$ over the possible energy losses of the particle and one obtains

$$\bar{A}(\omega)\,\mathrm{d}\omega = \mathrm{d}\Omega\cdot 16\pi^4 h^2 m^2\cdot G(\omega)\cdot h\,\mathrm{d}\omega\int_{h\omega}^{E}\frac{|p_0'|}{|p_0|}|(p_0'|V|p_0)|^2$$
$$\cdot S(E - h\omega)\cdot C\,\mathrm{d}E. \tag{38a}$$

Here $E_0 = p_0^2/2m$ is the initial energy of the particle. As we shall see, in some cases $S(E)$ has a sharp maximum in the neighborhood of $E = 0$. In that case, it is allowable to insert for p_0'

$$(p_0^2 - 2mh\omega)^{1/2},$$

and to pull the first two factors in (38a) out of the integral, and also to hold the quantity C constant in the integration. This quantity C occurs in the function S in accordance with (32) and (33). In this case, it is useful to bear in mind the relation

$$\int_0^\infty S(E)\,\mathrm{d}E = 1 \text{ for fixed } |p_0'|. \tag{40}$$

According to the definition (31) of S, this relation is identical to Equation (21a) for $n = 0$. One can also easily verify that expression (36) satisfies this condition.

In the following section, the result summarized in Equations (30) and (38) shall now be discussed in more detail in special cases of physical interest.

§4 Discussion of two limiting cases

A $C \gg 1$

If we exclude very small deflection angles of the particles, $(\vec{p}_0 - \vec{p}'_0)^2$ becomes of order p_0^2, so that the presupposed inequality assumes the form

$$\frac{e^2}{hc} \cdot \frac{E_0}{mc^2} \gg 1 \tag{41}$$

if

$$E_0 = \frac{p_0^2}{2m} = h\omega_0 \tag{42}$$

denotes the initial energy of the particle, and ω_0 denotes the associated upper limit of the brems-spectrum. Together with the presupposition (I)

$$\frac{e^2 \omega_1}{mc^3} \ll 1 \tag{I}$$

there immediately follows from this

$$\omega_1 \ll \omega_0 \tag{43}$$

which means that the quantum edge of the spectrum is here really never reached at all, since the spectrum is cut off much sooner due to the finite size of the charged body.

With the special choice $G(\omega) = \exp -\omega/\omega_1$, one immediately sees from (36) that, for fixed C, the function $S(E)$ has a maximum at $E - (C-1)h\omega_1 \sim h\omega_1 C$. For large C, this maximum is very steep, for application of Stirling's formula yields

$$h\omega_1 S(E) \simeq \frac{1}{(2\pi C)^{1/2}} \exp\left[-\frac{1}{2}\frac{(E - h\omega_1 C)^2}{(h\omega_1)^2 \cdot C}\right]; \left(\frac{E - h\omega_1 C}{h\omega_1 C}\right)^2 = \frac{1}{C}. \tag{44}$$

This means that the relative margin of E, where $S(E)$ differs considerably from zero, becomes of the order $C^{-1/2}$ which is very small for large C. Considering (40), one can thus to a first approximation set

$$S(E) = \delta(E - h\omega_1 C). \tag{45}$$

We now show that this result for large C is independent of the special form of the cutoff. From (32) and (33) it is easily seen that, for large C, only the values of f for small $\alpha h\omega_1$ are relevant. If, in (33) one now expands the exponential function in the integrand in terms of α, the term linear in α yields

$$f(\alpha) = C \cdot (-\mathrm{i}\alpha h)\int^{\infty} G(\omega)\,\mathrm{d}\omega = -C\mathrm{i}\alpha h\omega_1$$

with the definition (7) of ω_1, and inserting this into (32) there again follows (45) for $S(E)$. Including the terms quadratic in α in the series expansion of $f(\alpha)$ then again yields a Gaussian error distribution of $S(E)$ with a relative margin of order $C^{-1/2}$ about the maximum $E - Ch\omega_1$.

Our result means that, in the limiting case under consideration, for a given angular deflection of the body, the energy loss is given by

$$E = Ch\omega_1 = \frac{2}{3\pi} \cdot \frac{e^2}{c^3}(\vec{v} - \vec{v}')^2 \cdot \omega_1, \tag{46}$$

if $\vec{v} = \vec{p}_0/m$ and $\vec{v}' = \vec{p}'_0/m$ denote the initial and final velocities of the particle. Since, according to (I), this is very small compared to the initial energy E_0, we can, in (46), set the magnitude of $|v'|$ equal to $|v|$ and we can write

$$E = \frac{2}{3\pi}\frac{e^2}{c^3}v^2 \cdot 2 \cdot (1 - \cos\theta)\omega_1. \tag{46a}$$

Here, θ is the deflection angle of the particle.[14]

By inserting (45) into (38a), one obtains for the spectral emission coefficient

$$\begin{aligned}&\bar{A}(\omega)\,\mathrm{d}\omega\\ &= \mathrm{d}\Omega \cdot 16\pi^4 h^2 m^2 \cdot \frac{2}{3\pi}\frac{e^2}{c^2} \cdot (\vec{v} - \vec{v}')^2 \frac{|p'_0|}{|p_0|}\\ &\cdot |(p'_0|V|p_0)|^2 G(\omega)\,\mathrm{d}\omega.\end{aligned} \tag{47}$$

Here, p'_0 is first of all determined from (39) as

$$E = h\omega + \frac{2}{3}\frac{e^2}{c^3}(v - v')^2 \cdot \frac{\omega_1}{\pi}.$$

Because $C \gg 1$, however, h is always negligible here compared to the second terms, unless $\omega \gg \omega_1$. But in that case, $A(\omega)$ vanishes anyway (or becomes very small) because of the cutoff factor. Secondly, because of (I), the second term in E is always small compared to E_0, so that, in (47), one can set

$$|p'_0| \sim |p_0|$$

with sufficient accuracy. As long as $\omega \ll \omega_1$, the spectral emission coefficient, for a given deflection angle of the particle, therefore is here independent of ω and, for $\omega \sim \omega_1$, becomes simply proportional to the cutoff factor $G(\omega)$.

The behavior of the charged body (energy loss and emitted spectrum) for a given angular deflection is here precisely the same *as would follow from classical electrodynamics if the velocity change of the body from* $\vec{v}$ *to* $\vec{v}'$ *were to take place 'suddenly', i.e. in a time that is short compared to* $1/\omega_1$.[15] The total cross section (integration over $\mathrm{d}E$) for the deflection of the body into a given solid angle, on the other hand, according to (30), (40), and (45) is the same as in the first Born approximation of nonradiative wave mechanics.

It is noteworthy that in this case, where the deflection of the body cannot be treated according to classical mechanics, the classically calculated 'collision time' also plays no role. On the other hand, it would be very desirable, in the inverse case of a classical force field (in which naturally the Born approximation fails), to investigate the transition to the limit of classical electrodynamics in the case $C \gg 1$. Here, the value of ω_1 (dimension of the body) would have to drop out of the result, if the vectorial velocity change in the time $1/\omega_1$ is always small compared to the particular velocity involved.

[14] By separate calculations, we have convinced ourselves that the terms of the Hamiltonian operator (I) which are proportional to $\vec{A}^2$ and which have been neglected here cause only corrections of higher order in the quantity $e^2\omega_1/mc^3$, which is assumed to be small, even in the limiting case $C \gg 1$, which is being considered here.

[15] In this connection compare also N. Bohr and L. Rosenfeld, *Det Kgl. Danske Vidensk. Selsk, Math. fyr. Meddel.* **XII**, 8, 1933.

B $C \ll 1$

This case is especially interesting since it allows a comparison with ordinary radiation theory, in which one expands in powers of $(e^2/hc)\cdot(v_0^2/c^2)$. As already mentioned in the introduction, in the case of weak force fields, this leads to the result that not only elastic scattering occurs with the cross section

$$dq = d\Omega \cdot 16\pi^3 h^2 m^2 |(p_0'|V|p_0)|^2 \tag{48a}$$

but also inelastic scattering whose cross section is given by[16]

$$dq = d\Omega \cdot 16\pi^4 h^2 m^2 \frac{|p_0'|}{|p_0|} |(p_0'|V|p_0)|^2 \cdot \frac{2}{3\pi} \cdot \frac{e^2}{hc} \frac{(\vec{p}_0' - \vec{p}_0)^2}{m^2c^2} \cdot \frac{dE}{E} \tag{48b}$$

if the energy loss of the particle lies between E and $E + dE$. The integral over dq becomes logarithmically infinite at $E = 0$ ('infrared catastrophe').

Since this theory considers only such processes in which only a single light quantum is emitted, the spectral emission coefficient turns out to be

$$\bar{A}(\omega)\, d\omega = d\Omega \cdot 16\pi^4 h^2 m^2 \frac{|p_0'|}{|p_0|} |(p_0'|V|p_0)|^2 \cdot \frac{2}{3\pi} \frac{e^2}{c^3} \frac{(\vec{p}_0' - \vec{p}_0)^2}{m^2} d\omega. \tag{48c}$$

If we now presuppose $G(\omega) = \exp -\omega/\omega_1$ for the sake of simplicity, the result of the theory being developed here can be read off directly from (30), (36) and (38), if we also consider that, since $C \ll 1$, we can set $\Gamma(C) = (1/C)\Gamma(C+1) \sim 1/C$

$$S\,dE = C\frac{dE}{E}\left(\frac{E}{h\omega_1}\right)^C \exp(-E/h\omega_1) = d\left(\frac{E}{h\omega_1}\right)^C \exp(-E/h\omega_1). \tag{49}$$

In particular, for the cross section there follows

$$dq = d\Omega \cdot 16\pi^4 h^2 m^2 \frac{|p_0'|}{|p_0|} |(p_0'|V|p_0)|^2 C \frac{dE}{E}\left(\frac{E}{h\omega_1}\right)^C \exp - E/h\omega_1. \tag{50}$$

As already mentioned, there is no strictly elastic scattering in this theory. The integral of (50) between the limits 0 and E is now finite because of the additional factor E^C and, up to terms of order C, has the value

$$\left(\int dq\right)_0^E = d\Omega 16\pi^4 h^2 m^2 |(p_0'|V|p_0)|^2 \left(\frac{E}{h\omega_1}\right)^C. \tag{51}$$

The neglected terms of order C come from $|p_0'|$ and C depending on the integration variable E, while here the respective values for $|p_0'| = |p_0|$ were inserted. If

$$-C \lg\frac{E}{h\omega_1} \ll 1 \text{ or } \lg h\omega_1 - \lg E \ll \frac{1}{C} \tag{52}$$

then one can set $(E/h\omega_1)^C \sim 1$, *and (51) agrees with the cross section of the old theory for elastic scattering, which is given by (48a). Likewise, under the presupposition in (52), the more general expression (50) agrees with (48b) – apart from the cutoff factor* $\exp(-E/h\omega_1)$. *As regards the behavior of the particle, (52) is thus the validity condition*

[16] Compare e.g. W. Heitler, *The Quantum Theory of Radiation*, p. 166, Equation (17). Corresponding to the point charge, we here set $G(\omega) = 1$.

for the old theory. Only a very narrow region in the neighborhood of $E = 0$ is thus excluded from the application range of the old theory. The size of this region, on the other hand, depends significantly on the value ω_1 of the cutoff frequency.

For the emission coefficient one obtains from (38), by series expansion of $|p_0'|$ at the point $E = h\omega$, and neglecting quantities of order C^2, the result

$$A(\omega)\,\mathrm{d}\omega = \mathrm{d}\Omega \cdot 16\pi^4 h^2 m^2 \frac{|p_0'|}{|p_0|} |(p_0'|V|p_0)|^2 \cdot \frac{2}{3\pi} \cdot \frac{e^2}{c^3} \frac{(\vec{p}_0 - \vec{p}_0')^2}{m^2}$$

$$\cdot \exp(-\omega/\omega_1) \int_0^{\omega_0 - \omega/\omega_1} C \cdot x^{C-1} \exp -x \,\mathrm{d}x \tag{53}$$

where one must insert $|p_0'| = (p_0^2 - 2mh\omega)^{1/2}$.

Up to now, no assumptions were made concerning ω_0/ω_1, and the derived formulas claim to be valid for arbitrary values of this ratio. In particular, if $\omega_1 \ll \omega_0$ (large body), the integral in (53) can simply be set equal to 1 for $\omega_0 - \omega \gg \omega_1$. While this case offers nothing surprising, we now wish to presuppose expressly

$$\omega_0 \ll \omega_1 \tag{54}$$

so that one can always assume $\exp -E/h\omega_1 \sim 1$ and $\exp -\omega/\omega_1 \sim 1$, $\exp -\omega_0/\omega_1 \sim 1$. If one then consistently neglects terms of higher order in C, one has

$$\mathrm{d}q = \mathrm{d}\Omega \cdot 16\pi^4 h^2 m^2 \frac{|p_0'|}{|p_0|} |(p_0'|V|p_0)|^2 C \cdot \left(\frac{E}{h\omega_1}\right)^C \frac{\mathrm{d}E}{E} \tag{50a}$$

$$\bar{A}(\omega)\,\mathrm{d}\omega = \mathrm{d}\Omega \cdot 16\pi^4 \omega h^2 m^2 \frac{|p_0'|}{|p_0|} |(p_0'|V|p_0)|^2$$

$$\cdot \frac{2}{3\pi} \frac{e^2}{c^3} \frac{(\vec{p}_0 - \vec{p}_0')^2}{m^2} \left(\frac{\omega_0 - \omega}{\omega_1}\right)^C . \tag{53a}$$

The limits of validity of ordinary radiation theory are now as before determined by (52). The latter formula contains the result that, in the immediate neighborhood of the edge, the spectral intensity falls to zero more rapidly than in ordinary radiation theory.[17]

The surprising feature in this result is that, even if $\omega < \omega_0 \ll \omega_1$, the value of ω_1 does not drop out of the result but essentially determines the limit of validity of the ordinary

[17] I would here also like to present a generalization of the results valid under presupposition (54) for an arbitrary type of cutoff. In this case, the aim is to determine the function $f(\alpha)$, defined by (33), for large values of the argument $\alpha h\omega_1$. The generalization of (35) is

$$f(\alpha) = -C\left\{\lg(\alpha h\omega_1) + \mathrm{i}\frac{\pi}{2} + \gamma\right\} + \ldots \text{ for } \alpha h\omega_1 \gg 1, \tag{$*$}$$

where γ is a constant of order of magnitude 1. This formula can be derived, for example, from the corresponding formula for $f'(\alpha)$. For small $E/h\omega_1$, there then follows from (32)

$$S\,\mathrm{d}E = \frac{\mathrm{d}E}{E}\left(\frac{E}{\mathrm{e}^\gamma \cdot h\omega_1}\right)^C \frac{1}{\Gamma(C)} \sim C\frac{\mathrm{d}E}{E}\left(\frac{E}{\mathrm{e}^\gamma h\omega_1}\right)^C \tag{49'}$$

for small C and $E/h\omega_1$. This means that ω_1 should always be replaced by $\omega_1 \cdot \mathrm{e}^\gamma$. The following variant of the summation method of §3 also leads to the same result. Let us set $S = (\partial S'/\partial E)_{p_0}$ (differentiation for fixed p_0') with $S' = \sum_{h\Sigma_s n_s\omega_1 < E} (n_s) \prod_s (1/n_s!) w^{n_s} \exp(-w_s)$. For small C, it is allowable to add the factor $\exp -h\sum_s n_s\omega_s/E$ instead of the auxiliary condition $h\sum n_s\omega_1 < E$, and the sum over all n_s. The result is $S' = \exp(\varphi(1/E))$; $\varphi(\alpha) = f(-\mathrm{i}\alpha) = c\prod_0^\infty (\mathrm{d}\omega/\omega)(\exp(-\alpha h\omega) - 1)G(\omega)$. From ($*$) there again follows (49′) for $E/h\omega_1 \ll 1$. For a sharp cutoff [$G(\omega) = 1$ for $\omega < \omega_1$ and $G(\omega) = 0$ for $\omega > \omega_1$], γ is the Euler constant.

theory. It is therefore important to determine the frequencies of the sum S, as defined by (31), from which the dependency of its value on ω_1 originates. If (for fixed p_0') we form

$$S' = \int_0^E S\,\mathrm{d}E = \sum_{h\sum_s n_s\omega_s < E} (n_s)\prod_s \frac{1}{n_s!} w^{n_s} \exp - w$$

it is readily seen that this is caused *by the term with $n_s = 0$ for $\omega > E/h$*. If we here assume a sharp cutoff, it is easily seen that

$$\prod_{\omega_s > E/h} \exp - w = \exp\left\{-C\int_{E/h}^{\omega_1} \frac{\mathrm{d}\omega}{\omega}\right\} = \left(\frac{E}{h\omega_1}\right)^C$$

yields the main term.[18] Furthermore, we obtain an upper limit for S' if we sum over *all* n_s for $\omega_s < E/h$ (which always yields the value 1 in $\prod_s$) instead of adhering to the inequality $\sum_s n_s\omega_s < E/h$. Consequently[19] we have

$$S' < \left(\frac{E}{h\omega_1}\right)^C.$$

It should also be pointed out that the formal expansion of (50a) in powers of C, according to

$$\left(\frac{E}{h\omega_1}\right)^C = 1 + C\,\mathrm{lg}\left(\frac{E}{h\omega_1}\right) + \ldots$$

(where naturally the integrability of $\mathrm{d}q$ at $E = 0$ is again lost) must agree with the results of the higher approximations of ordinary radiation theory, in which one expands in terms of e^2/hc. In fact, these higher approximations diverge for point charges.

The dependence of the result on ω_1 on the one hand prevents application to real electrons. Of course, one could think about replacing ω_1 by a quantity of the order mc^2/h, in contradiction to presupposition (II), but such a procedure would be purely speculative. On the other hand, this dependency seems to show that the finer features of the quantum electrodynamics of spatially extended charges, in the case $\frac{1}{2}mv_0^2 \ll h\omega_1$, cannot be brought into direct connection with considerations based on the correspondence principle.

Zürich, Physical Institute of the E. T. H.

[18] According to the remark on [p. 232], neglect of the retardation is permitted precisely for $n_s = 0$.

[19] Compare the factor e^γ in [footnote 17].

10 The universal length appearing in the theory of elementary particles

W. HEISENBERG

Annalen der Physik, Ser. 5, 32: 20–33 (1938). Received 13 January 1938.

This issue celebrates the creator of quantum theory. More than any other discovery of the new physics, this has led to a general change and clarification of our physical world-picture. This occasion may serve as an excuse if the following considerations deal more with general and largely known relationships than would otherwise perhaps seem allowable in the context of an individual research.

Sometime ago, the author pointed out[1] that a consistent application of the Fermi theory of β-decay forces the conclusion that, when an extraordinarily energetic cosmic-ray particle collides with another particle, many particles may possibly be created in explosive fashion. This process should happen if the wavelengths of the colliding particles in the center-of-mass system fall below a certain critical length. This length enters the Fermi theory of β-decay as a universal length of the order of the classical electron radius

$$\frac{e^2}{mc^2} = r_0 = 2.81 \times 10^{-13} \text{ cm.}$$

The existence of such explosions at that time seemed very probable because of experiments on cosmic-ray showers and Hoffmann collisions. In the meantime, however, the theories of Carlson and Oppenheimer[2], as well as Bhabba and Heitler[3], have shown that a major portion of these phenomena can be interpreted simply by quantum electrodynamics through so-called cascade formation. Furthermore, Yukawa[4] and Wentzel[5] have proposed different theories of β-decay. Apparently, these theories represent experiments just as well as the Fermi theory does and furthermore they promise[6] to link β-decay with the existence of particles recently discovered by Neddermeyer and Anderson.[7] Consequently, the thesis that a universal natural constant with a length dimension becomes visible in the occurrence of explosions during the collision of short wavelength particles certainly does not stand on a firm foundation. Consequently, we shall below discuss the arguments individually, which suggest that nuclear physics and the theory of cosmic radiation must take into account the universal length of the order r_0, and that this length is related to the possible existence of the explosions.

1 W. Heisenberg, *Ztschr. f. Phys.* **101**, p. 533, 1936.
2 J. F. Carlson and J. R. Oppenheimer, *Phys. Rev.* **51**, p. 220, 1937.
3 H. Bhabba and W. Heitler, *Proc. Roy. Soc.* A **159**, p. 432, 1937.
4 H. Yukawa, *Proc. Phys. Math. Soc. Japan* **17**, p. 48, 1935.
5 G. Wentzel, *Ztschr. f. Phys.* **104**, p. 34, 1937; **105**, p. 738, 1937.
6 Compare especially J. R. Oppenheimer and R. Serber, *Phys. Rev.* **51**, p. 1113, 1937; E. C. G. Stückelberg, *Phys. Rev.* **52**, p. 42, 1937; H. Yukawa and S. Sakata, *Proc. Phys. Math. Soc. Japan* **19**, p. 1084, 1937; N. Kemmer, *Nature* **141**, p. 116, 1938; H. Bhabba, *Nature* **141**, p. 117, 1938.
7 S. Neddermeyer and C. P. Anderson, *Phys. Rev.* **51**, p. 884, 1937; compare also J. C. Street and E. C. Stevenson, *Bull. Amer. Phys. Soc.* **12**, pp. 2, 13, 1937.

I The constants c and $\hbar$

The development of the new physics shows that the effectiveness of a fundamental natural constant is first noticed through the contradictions which appear in the consistent development of comprehensive theories that appear thoroughly secure experimentally, especially in the conjoining of two such theories. Thus, at the turn of the century, Newtonian mechanics and the Maxwell theory were considered secure possessions of physics. However, considerable contradictions appeared in the attempt to combine the two theories and to work out the electrodynamics of moving bodies. These contradictions can only be removed by recognizing the velocity of light c as a universal natural constant of the most general significance, which must be taken into account in the formulation of every physical law. In similar fashion, the statistical theory of heat, as worked out by Gibbs and Boltzmann, must be regarded as definitive. When applying this theory to radiation problems, that is when joining the Maxwell theory and the theory of heat, very severe contradictions appear, however: For radiation in thermal equilibrium, divergent results were obtained in the Rayleigh–Jeans law. Only Planck's discovery showed that, when conjoining these theories, one must take into account a universal natural constant of the dimension of action.

When the general laws associated with the constants c and $\hbar$ were clearly understood, it was recognized that what was involved was not really a correction of theories that previously were regarded as secure. The previous theories continued to have standing as intuitive limiting cases in which the velocity of light can be regarded as very large and the action quantum as very small. The constants c and $\hbar$ rather designate the limits in whose proximity our intuitive concepts can no longer be used without misgivings. This state of affairs has often been expressed by the statement that the earlier theories emerged from relativity theory and from quantum theory through the limiting process $c \to \infty$, $\hbar \to 0$. However, this formulation is not quite unproblematical because it can be correct only if certain quantities are held constant during this limiting process (e.g., in the transition from quantum mechanics to classical mechanics, the masses and charges of the elementary particles). In a definitive theory, however, these quantities would be determined from the few universal constants of physics, and a change of the magnitude of the universal constants could change nothing at all in the form of the physical laws. Incidentally, the opposite limiting process $\hbar \to \infty$ or $c \to 0$, while holding the above quantities constant, leads to meaningless results. It therefore appears more correct to stay with the first formulation and to designate $\hbar$ and c simply as the limits which are set to the application of intuitive concepts.[8]

Incidentally, the constants $\hbar$ and c differ from other, less fundamental, universal constants, in virtue of this property. For example, the Boltzmann constant k is connected with the arbitrary specification of the temperature scale and could be calculated from a theory of the states of aggregation of water, or it could be omitted entirely when measuring temperature on an energy scale. Even a universal constant like the mass of the proton in no way has such a basic significance as $\hbar$ or c; because the mass concept undoubtedly has a simple specifiable meaning for even smaller masses; e.g. the mass of a light quantum can still be measured by way of the light pressure. It therefore appears

[8] In this connection compare especially N. Bohr, *Atomic Theory and the Description of Nature*, Berlin, Springer, 1931.

suitable to distinguish constants with such fundamental properties as $\hbar$ and c as 'universal constant of the first kind' as opposed to other ones.

These universal constants of the first kind, if one may generalize from the already known examples $\hbar$ and c, are always linked with a very general property of natural laws, namely a type of invariance property; they impose general requirements on the form of any physical law. Special relativity requires the invariance of all physical laws with respect to Lorentz transformations; quantum mechanics requires the commutation relations between canonically conjugate variables, the existence of probability amplitudes, and invariance of laws with respect to rotations in Hilbert space.

Apart from gravitational phenomena, which scarcely appear to play a role in atomic physics, no other universal constants of the first kind, apart from $\hbar$ and c, are currently known in microphysics.

II The universal length

But now it is clear in advance that, in atomic physics, there must also exit another 'universal constant of the first kind' having the dimension of a length or a mass. A length or mass cannot be formed dimensionally from the constants $\hbar$ and c. Consequently, the mass of the elementary particles and the dimensions of the atoms and of the atomic nuclei must be specified by another universal constant. Now a universal mass m can be formed from a universal length r_0 and inversely a length can be formed from a mass through the relation $m = \hbar/cr_0$, and consequently it does not seem to matter whether one speaks of a universal length or mass. Still, the physical significance of this constant is expressed more clearly by introducing it as a universal length. For then it again signifies a limit to the application of intuitive concepts: The concept of a length can be used without restriction only for distances which are large compared to the universal length. A similar physical interpretation of a universal mass, on the other hand, is not possible in such simple fashion, for larger and smaller masses are accessible to measurement with any accuracy. Consequently, we wish to introduce the new constant as a length and we wish to assign to it the value $r_0 = e^2/mc^2 = 2.81 \times 10^{-13}$ cm. This latter specification is evidently arbitrary within certain limits, just as one can arbitrarily introduce h or $\hbar$ as the universal constant of action. Only the complete theory will show which specification is the most suitable and naturally it is improbable that the precise value of r_0 will prove to be the most suitable; however, as further considerations will show, it has in any case the right order of magnitude. The specification of r_0 also should *not* mean that the universal length is supposed to be directly connected with the question of the electron charge.

We shall explain below that – as has already been said in various places – the contradictions which everywhere have up to now occurred in quantum electrodynamics, in the theory of β-decay, and in the theory of nuclear forces, vanish if one notes the restrictions which are prescribed by the universal length r_0; further that the length r_0 must play a decisive role in the theory of elementary particles, and that the restrictions of the possibilities of measurement, due to r_0, perhaps can be made easily intelligible through the existence of explosions.

III The divergences

If one applies the methods of quantum theory to a relativistically invariant wave theory which also includes interactions between the waves (i.e. nonlinear terms in the wave equation), one obtains divergent results, as has been frequently noted. The reason for this is that relativistic invariance requires a 'short-range theory', in which the interaction is caused by the propagation speed of a wave at a point being determined by the amplitude of another wave at this point. Because of the infinitely many degrees of freedom of the continuum, i.e. because of the possibility of waves of arbitrarily small wavelength, however, the eigenvalues of a wave amplitude at a particular point become infinite. This contradiction – which greatly resembles the contradiction in the Rayleigh–Jeans law – now evidently does not mean that the relativistic wave theory or the quantum theory are wrong or need improvement, but rather it points to the idea that, when quantum theory and relativistic wave theory are conjoined, account must be taken of a universal constant having the dimension of a length. In fact, many authors have availed themselves of the remedy that they artificially made convergent or broke off the divergent integrals at a length of order r_0 (or at the corresponding momenta), thus achieving reasonable results. But such a termination cannot generally be implemented in a relativistically invariant fashion and naturally is to be regarded only as a very provisional emergency remedy. Indeed, in a definitive theory, the qualitatively new physical phenomena which occur at length of order r_0 (and which naturally satisfy relativistic invariance requirements) would instead have to be taken into account correctly, which by itself would lead to convergence of the integrals.

Taking into account these considerations, the question of the self-energy of the electron occupies a special position. Even in classical theory, this was regarded as a proof of the finite electron radius. This familiar classical substantiation for the radius r_0 undoubtedly cannot be taken over into quantum theory. The reason is that the charge of the electron is smaller than the dimensionally corresponding quantity $(\hbar c)^{1/2}$; consequently the non-intuitive features in the description of the electron field, which are due to quantum theory, cut the ground from under the above-mentioned classical consideration.[9] Perhaps one might have been able to hope that a quantum theory of the electron and its surrounding field exists in which no self-energy occurs at all, that is the electron has no rest mass. Such a theory would to a certain extent emerge from the correct theory through the limiting process $r_0 \rightarrow \infty$, but appears just as senseless as the previously discussed limiting process $\hbar \rightarrow \infty$ in quantum theory. However, as soon as a finite electron rest mass is to be derived from a theory, this theory must contain not only $\hbar$ and c but also a universal length r_0, that is elements which have nothing to do with electrodynamics and quantum theory. For this reason, it seems improbable that a theory of the Sommerfeld fine-structure constant $e^2/\hbar c$ can be found until the new features in the description of nature, due to r_0, have been clarified, features which at first have no connection at all with the question of the electronic charge.

On this occasion, we can perhaps briefly touch on the problem of gravitational forces which otherwise are not a subject of this discussion. The mutual attraction of two light quanta can be compared with the above-mentioned electric interaction between two

[9] Compare N. Bohr and L. Rosenfeld, *Dansk Vid. Selsk. math. phys. Medd.* **12**, p. 8, 1933.

electrons, and one can ask about the gravitational self-energy of the light quanta.[10] The fact that the light quanta have no rest mass here first of all suggests the idea that, in this problem, r_0 perhaps can play no role. However, it turns out that – in contrast to the electrical analog – the gravitational constant γ together with $\hbar$ and c itself characterizes a length: $l = (\hbar\gamma/c^3)^{1/2} = 4 \times 10^{-33}$ cm. The fact that this length is much smaller than r_0 gives us the right of initially neglecting, in the description of nature, the non-intuitive features due to gravitation, since these features – at least in atomic physics – are completely submerged in the much coarser non-intuitive features which originate from the universal constant r_0. For these reasons, it should indeed be scarcely possible to subsume the electric and gravitational phenomena into the rest of physics before the problems associated with the length r_0 have been solved.

The discussion of questions associated with r_0 thus appears to be the most urgent task. To treat these questions, one will primarily have to study the phenomena of nuclear physics and of cosmic radiation where one can to a first approximation neglect electrical and gravitational interactions.

In this area of physics, it is primarily the theory of β-decay where the above-mentioned divergence difficulty expresses itself in the quantization of the wave fields. In particular, if one uses as a basis the Fermi theory of β-decay, the consistent application of quantum theory leads to divergences of such high degree that the results depend not only quantitatively but also qualitatively on the manner in which the divergent integrals have been artificially converted into convergent ones. Thus, for example, the calculations of v. Weizsäcker,[11] Fierz[12] and others have shown that the forces between the elementary particles, which derive from the theory of β-decay, depend entirely on the type of 'cutoff' for small wavelengths. This fact has caused various researchers to doubt the qualitative consequences of the Fermi theory, which express themselves in the possibility of explosions, since a suitable cutoff at sufficiently long wavelengths can cause the above-mentioned consequences not to appear.[13] However, such a conclusion appears to me to be based on a misunderstanding. Indeed, the justification for a cutoff inversely can be derived only from the qualitatively new phenomena which appear at the critical wavelengths. If these qualitatively new phenomena are deleted, the cutoff method also loses any physical sense.

Yukawa and Wentzel (*loc. cit.*) have attempted to replace the Fermi theory of β-decay by another theory in which the process occurring during β-decay is composed of two elementary transitions. In this case, the divergences which occur are of lower degree than in the case of the Fermi theory. The theory of β-decay then has some similarity with ordinary radiation theory, where a charged particle with Bose statistics and with a mass of order $\hbar/cr_0$ replaces the light quantum. This particle can then perhaps be identified with the unstable particle suspected by Neddermeyer and Anderson (*loc. cit.*). The question whether, in this theory, explosions are to be expected at energies prevailing in cosmic radiation, depends on whether the quantity[14] $g^2/\hbar c$,

[10] Compare L. Rosenfeld, *Ztschr. f. Phys.* **65**, p. 589, 1930.
[11] C. F. v. Weizsäcker, *Ztschr. f. Phys.* **102**, p. 572, 1936.
[12] M. Fierz, *Ztschr. f. Phys.* **104**, p. 553, 1937.
[13] G. Nordheim, L. W. Nordheim, J. R. Oppenheimer, and R. Serber, *Phys. Rev.* **51**, p. 1037, 1937.
[14] H. Yukawa and S. Sakata, *loc. cit.*, p. 1090.

which corresponds to the Sommerfeld fine structure constant, can be regarded as small compared to unity when calculating the cross sections. Depending on the mass of the Yukawa particles, its value lies between 1/10 and 1. The consequences of this theory for the question of multiple processes can therefore not be seen clearly for the time being. In the limiting case $g^2/\hbar c \ll 1$, multiple processes become improbable; as a consequence, the theory does indeed continue to resemble radiation theory. However, it also still requires a 'cutoff method' which it must obtain from other phenomena, which are unknown and which are not contained within the theory. On the other hand, for values $g^2/\hbar c \sim 1$, this theory, too, leads to the possibility of explosions. But then it loses its resemblance to radiation theory, because an expansion in terms of $g^2/\hbar c$ would then make no sense.[15] In any case, what is involved here – just as in the Fermi theory – is surely only a correspondence-like analog to a definitive theory, in which the length r_0 must occur at an essential point. In the past, for instance, the example of the Uhlenbeck–Goudsmit theory of spin has shown very clearly how fruitful such a correspondence-like analog can be on the other hand – even in an area in which the non-intuitive features already play an essential role.

IV The theory of elementary particles

The fact that present theories of β-decay can only have the character of a correspondence-like analog seems to appear most clearly from the circumstance that, in these theories, the masses of the elementary particles occur only as special universal constants. In the definitive theory, these many different rest masses of the neutron, proton, electron, neutrino, and the new unstable particles would have to be derived in similar fashion from the constant r_0, such as, for instance, the terms of the hydrogen atom from the Rydberg constant. Now surely there will be large areas of physics in which the masses of the elementary particles can be regarded simply as fixed parameters, and in which the theory of these masses can be postponed until later; in particular, this will be the case wherever energy transformations are involved which, in the center-of-mass system, are small compared to the critical energy $\hbar c/r_0$. Within this region belongs not only the entire physics of the atomic shells but also ordinary nuclear physics and the theory of β-decay. Only if one tries to link the theory of β-decay with the theory of nuclear forces, or if one wishes to apply it to problems of cosmic radiation, will one need to make statements about processes with large energy transformation. In such statements, one will no longer be able to avoid dealing with the theory of elementary particles. One must thus conclude that all attempts to link the β-decay theory with nuclear forces, with the magnetic moment of the elementary particles, and with processes of cosmic radiation, can have only a very provisional character, as long as the masses of the elementary particles appear as independent constants in such theories. For, if processes are discussed where particles arise with a rest mass of order $\hbar c/r_0$, one will necessarily have to pay attention to qualitatively new phenomena which occur at

[15] If the results of B. Kockel, *Ztschr. f. Phys.* **107**, p. 153, 1937, may be transferred to the Yukawa–Wentzel theory and may be generalized, multiple processes beginning at about 10^8 eV would already be the rule at a value $g^2/\hbar c = 1/10$. Consequently, already for $g^2/\hbar c = 1/10$, the convergence of an expansion in powers of $g^2/\hbar c$ is very questionable.

the length r_0. These phenomena on the one hand must specify the elementary masses and on the other hand must substantiate the elimination of the divergences discussed in Section III.

V New phenomena due to the universal length r_0

Now, what are the new phenomena which occur at distances or wavelengths of the order r_0? As long as only the motion of individual particles is involved, the constant r_0, on account of relativistic invariance requirements, can express itself only in the appearance of a rest mass. Naturally it makes no difference at all whether this single particle has an energy that is large or small compared to the critical energy $\hbar c/r_0$, since the energy depends on the reference system of the observer.

However, when two particles interact, it will be essential for the further physical events whether the kinetic energy of the particles in their center-of-mass system is large or small compared to $\hbar c/r_0$ when these particles collide. In the case of *small* energies, as the experiments of nuclear physics show, the behavior of the particles can be regarded as if a force were to act between them which assumes noticeable values only at distances of order r_0 (in the center-of-mass system). These interaction energies of the order $\hbar c/r_0$ and range r_0 are, so to speak, the first characteristic feature of the constant r_0. It therefore also appears questionable to what extent it is suitable to regard these forces as being from the β-decay forces. In reality, both the nuclear forces and the β-decay forces form a unit, and it will scarcely be possible to speak of primary and of derived effects. For similar reasons, it surely could also be assumed that forces over the approximate range r_0 act between *all* types of elementary particles; an exception is forced, at most, by particles whose rest mass is very much smaller than $\hbar c/r_0$ (electrons, neutrinos, light quanta); with these particles, the ‘nuclear forces’ will perhaps also be especially weak.

Compared to what is known about the interaction of particles with small kinetic energy, very much less is known about the interaction of two particles which, during their collision, have an energy in the center-of-mass system that is large compared to $\hbar c/r_0$. Evidently, the processes which take place here must be closely connected with the non-intuitive features which are introduced into physics through the constant r_0; in a manner similar to how the behavior of an electron, e.g. in the normal state of the hydrogen atom, shows especially clearly the characteristic non-intuitive features of quantum theory.

Now perhaps it cannot be expected that one could survey all the possibilities for the expression of non-intuitive features originating from r_0, during the collision of high-energy particles. But the one possibility which is suggested by the Fermi theory shall still be discussed in detail.

It can be assumed that, when two particles collide, whose energy in the center-of-mass system is large compared to $\hbar c/r_0$, this energy is generally in one stroke divided into many elementary particles. This assumption of explosions is derived as a conclusion from the Fermi theory of β-decay. But quite independent of this theory, it is a logical possibility which satisfies all the invariance requirements originating from the theory of relativity and from quantum theory.

The creation of new elementary particles during the collision of two high-energy particles is indeed already suggested by the analogy with electrodynamics: The interaction of low-energy electrons is determined by the Coulomb force as is shown in detail in the theories of deceleration and ionization. However, according to Bethe and Heitler, radiation plays the principal role in the deflection of very high-energy electrons. This can be regarded as follows: In the deflection of an electron that moves at nearly the speed of light, its electric field cannot follow very easily on account of retardation; part of this field, as a light quantum, leaves the site where the deflection took place. As the theory of Bethe and Heitler shows, the light quantum can here frequently carry along a significant portion of the energy of the deflected particle. In similar fashion, it can be assumed that, when two elementary particles collide, which approach to distances of order r_0 at very high energy, the nuclear field cannot follow along so easily during the deflection, so that a portion of this field leaves the site of the collision in the form of elementary particles, which then again can carry along a large portion of the total energy.

This analogy also teaches that – if explosions occur at all – it must be expected that particles of *all* types can arise in the explosions. This assumption indeed is not suggested by the Fermi theory, since in this theory, for example, the newly discovered unstable particles do not occur; however, it appears to me a natural consequence from the physical foundations of the hypothesis of explosions. In particular, Neddermeyer–Anderson particles as well as protons and neutrons will frequently also arise during an explosion.

If one asks about the experimental verification of this hypothesis of explosions, one must search for features which make it possible to distinguish explosions reliably from cascades. An important feature first of all consists in the fact that the explosion will very frequently be coupled with a nuclear evaporation.[16] For, when a very high-energy particle collides with an electron at rest, the energy available in the center-of-mass system is always much smaller than when the same particle collides with a heavy particle at rest. When the high-energy particle collides with an electron, the energy in the center-of-mass system therefore will generally not be sufficient to form a rather large explosion, but it may well be sufficient in the case of a collision with a heavy particle. Since these particles are generally found in nuclei, the nucleus heated by the explosion will probably evaporate following the explosion. Many of the particles which arise during an explosion will no longer be able to form cascades (because of their rather large rest mass), and this can count as another characteristic of an explosion. Finally, the most important feature of the explosion remains its occurrence in a very thin layer. Some Wilson photographs by Fussell[17] very probably represent smaller explosions. A detailed analysis of experiments on Hoffmann collisions by Euler[18] also shows that explosions probably play a considerable role in these collisions. But here one must wait for further experiments.

If the explosions actually exist and if they represent processes that are actually

[16] In this connection also compare W. Heisenberg, *Ber. d. Sächs. Ak. d. Wiss.* **89**, p. 369, 1938, especially §6.

[17] L. Fussell, *Phys. Rev.* **51**, p. 1005, 1937. I am very indebted to Mr Fussell for transmitting to me several such photographs.

[18] H. Euler, to be published. [*Zeitschrift für Physik* 110: 692–716 (1938).]

characteristic of the constant r_0, they perhaps transmit an initial, as yet unclear, understanding of the non-intuitive features associated with the constant r_0.[19] These features will probably express themselves at first by the measurement of a length with a precision finer than the value r_0 being associated with difficulties. In quantum theory, it was the existence of matter waves or, more correctly, the side-by-side existence of wave and particle properties, which took care of the fact that those laws were not violated which were set by the uncertainty relations. In similar fashion, the explosions would be able to take care of the fact that positional measurements with a precision finer than r_0 would be impossible. For example, if one thinks about a positional measurement by a γ-ray microscope, one would have to use γ-rays with a wavelength less than r_0, that is light quanta with an energy greater than $\hbar c/r_0$, in order to attain the desired accuracy. But these light quanta would generally not be scattered at the object that is supposed to be observed, even if this object has a sufficiently large mass – it can here be at rest or in motion. Rather, explosions would be produced, in which the individual generated particles would have a wavelength of the order r_0. It is therefore impossible to image an object with a precision finer than r_0.

In recent years, two attempts have been made to build a universal length r_0 into the foundations of the formalism of atomic physics, in order to avoid the divergence difficulties of previous theories. In several papers, Born and Infeld[20] have attempted to modify the Maxwell theory in the region of small wavelengths so that the self-energy of the electron assumes a finite value. In a certain sense, these investigations do represent the exact fulfillment of the program of the Lorentz electron theory, since they take into account the finite rest mass of the electron in a relativistically invariant and consistent fashion. However, up to now it has not been possible to expand them into a quantum theory of the electromagnetic field. Also, they probably take too little account of the circumstance that the new phenomena, appearing at the length r_0, seem to have their root not in electrodynamics but in nuclear physics. A quite different attempt to remove the divergences has been undertaken by March,[21] who proposes to modify geometry at small lengths. Now such a modification of geometry corresponds to the suspicion that our intuitive concepts are applicable only down to lengths of order r_0. But the question is whether, in a formalism like that of March, too many concepts of previous physics are still being used uncritically; also, up to now it has not been possible to connect the March ideas with experiments in nuclear physics and in cosmic radiation.

If one thinks about the comprehensive changes which the formal representation of natural laws has undergone by understanding the constants c and $\hbar$, one will expect that the length r_0 will also force the formation of completely new concepts which will find an analog neither in quantum theory nor in relativity theory. In particular, it is conceivable that an invariance requirement exists here too, which can be formulated by means of the constant r_0, and which all natural laws must satisfy. Perhaps, in attempting to pursue these new conceptual structures, it may initially be advantageous to recall the fact that theoretical physics can only ever be concerned with the mathematical

[19] In this connection, I am indebted to Mr N Bohr for many instructive discussions.

[20] M. Born, *Proc. Roy. Soc.* (A) **143**, p. 410, 1933; M. Born and L. Infeld, *ibid.* **144**, p. 425, 1934; **147**, p. 522, 1934; **150**, p. 141, 1935.

[21] A. March, *Ztschr. f. Phys.* **104**, p. 93 and p. 161, 1936; **105**, p. 620, 1937; **106**, p. 49, 1937; **108**, p. 128, 1937.

linkage of observable quantities; for the time being, our only task is to find computational rules by means of which we can link the cross sections of cosmic radiation processes on the one hand among one another and on the other hand with other simple observational data. But, to accomplish such a program successfully, a considerable expansion of observational material available hitherto would also be a necessary presupposition.

Leipzig O 27, Bozener Weg 14.

11 The interaction between charged particles and the radiation field

PROF. H. A. KRAMERS

Nuovo Cimento NS, 15: 108–17, 1938. Presented at the Galvani Bicentenary Congress, Bologna, 18–21 October, 1937.

In a recent publication,[1] I have developed the basic formulas of the quantum theory of the interaction between the radiation field and charged particles. This development was in a form that differs somewhat from the usual presentations in the literature. The most consistent of these presentations goes back to the fact[2] that, in classical electron theory, equations of motion of charged, radiating particles with a finite size and the equations which refer to the electromagnetic field can be written in canonical form; nothing stands in the way of a formal quantization of these canonical equations. But what causes trouble is the fact that the concept of a radiating body contravenes the foundations of relativity theory. To this must be added that one encounters major and well-known difficulties if one investigates the transition to the limit where the extension of the particles tends to zero and where the electromagnetic mass increases more and more.

In the above-mentioned publication, I have tried to present the theory in such a fashion that the questions of the structure and the finite extension of the particles are not explicitly involved and that the quantity that is introduced as the 'particle mass' is from the very beginning the experimental mass. Indeed, I start from phenomena – and for the time being we speak purely classically – in which a charged particle moves in an external magnetic field and for which radiation and the radiation reaction can be neglected to a first approximation (quasi-stationary motion). This motion is governed by a Hamiltonian energy function which I designate as $H^{(\mathrm{mat})}$. The experimental mass m appears in this function: one can even assert that the use of the mass concept is defined by $H^{(\mathrm{mat})}$. The variables on which $H^{(\mathrm{mat})}$ depends on the one hand are the space coordinates x, y, z (vector $\vec{r}$) and the time, and on the other hand the components p_x, p_y, p_z of the momentum vector $\vec{p}$. Because of gauge invariance, $\vec{p}$ occurs in the combination $\vec{p} - (e/c)\vec{A}^{\mathrm{ext}}$, where $\vec{A}^{\mathrm{ext}}$ is the vector potential of the field at the position of the particle and consequently is a function of x, y, z, t.

We need not discuss the precise form of $H^{(\mathrm{mat})}$; but it should be expressly emphasized that the vector potential appearing in $H^{(\mathrm{mat})}$ and likewise the scalar potential Φ refer to the *external* electromagnetic field prevailing at the position of the particle. Without this concept of the external field, the use of $H^{(\mathrm{mat})}$ makes no sense. To obtain the total electromagnetic field, one must add to the external field still another field which increases as $1/r^2$ at small distances from the particle and which there behaves like the electromagnetic field of a point charge, which has the velocity of the particle.

[1] *Hand und Jahrbuch der Chemischen Physik* (*Manual and Yearbook of Chemical Physics*) I, Section 2, 'Quantum theory of the electron and of radiation', [esp.] §89, §90 (Leipzig 1938).
[2] Mr Pauli made express reference to this in his discussion about this lecture.

The following formula for the velocity of the particle (we write it down only for the x component) is also important for subsequent calculation:

$$v_\alpha = \frac{\partial H^{(\mathrm{mat})}}{\partial p_x} = -\frac{c}{e}\frac{\partial H^{(\mathrm{mat})}}{\partial A_x^{\mathrm{ext}}} \tag{1}$$

Now, to obtain a description of the effect of the particle on the radiation field, it is appropriate to decompose this field into its Fourier components. It thus becomes possible to obtain a formalism in an unforced fashion, in which the structure and extension of the particle do not appear explicitly: One can regard it formally as a point charge, and a finite extension only means that the equations which refer to Fourier components with very high wave numbers (with electrons about 10^{13} cm^{-1} and larger) will not be quite correct.

The total electromagnetic field $\vec{F} = \vec{E} + \mathrm{i}\vec{H}$ is first decomposed into a source-free component and into a curl-free component $\vec{F}_{\mathrm{long}}$. The latter is nothing more than the electric Coulomb field of the particle thought of as being at rest. In a Fourier decomposition, it consists only of longitudinal waves. The former component is now decomposed into clockwise and counterclockwise circular transverse waves, according to the following equation:

$$\vec{F}_{\mathrm{tr.}} = \Omega^{-1/2}\sum_\lambda (a_\lambda \vec{e}_\lambda \exp 2\pi \mathrm{i}\vec{\sigma}_\lambda\cdot\vec{r} + b_\lambda\vec{e}_\lambda \exp -2\pi\mathrm{i}\vec{\sigma}_\lambda\cdot\vec{r}) \tag{2}$$

(as regards the nomenclature, we refer to the cited publication). The following equations of motion now hold for the a_λ:

$$a_\lambda + 2\pi\mathrm{i}\nu_\lambda a_\lambda = -\frac{4\pi}{\sigma_\lambda^2\Omega^{1/2}}\, e(\vec{v}\vec{e}^*)\exp -2\pi\mathrm{i}\vec{\sigma}_\lambda\cdot\vec{r} \tag{3}$$

(and similarly for b_λ). These follow directly from the basic equations of classical electron theory if these are specialized to point charges. Let us now imagine that we know which part of $\vec{F}_{\mathrm{tr.}}$ should always be regarded as the external field:

$$\vec{F}_{\mathrm{tr}}^{\mathrm{ext}} = \Omega^{-1/2}\sum_\lambda (a'_\lambda \vec{e}_\lambda \exp 2\pi \mathrm{i}\vec{\sigma}_\lambda\cdot\vec{r} + b'_\lambda\vec{e}_\lambda \exp -2\pi\mathrm{i}\vec{\sigma}_\lambda\cdot\vec{r}) \tag{4}$$

Then $H^{(\mathrm{mat})}$ also becomes a function of the Fourier coefficients a'_λ, b'_λ, and (3) can be written in the following form, with the aid of (1):

$$\dot{a}_\lambda + 2\pi\mathrm{i}\nu_\lambda a_\lambda = -\mathrm{i}\frac{16\pi^2 c}{\sigma_\lambda}\frac{\partial H^{(\mathrm{mat})}}{\partial a_\lambda'^*} \tag{5}$$

(and similarly for b_λ).

The usual radiation theory, one might say, now starts from the idea that the difference between the external field and the total field can be neglected, that is the $a_\lambda'^*$ in (5) are replaced by the a'_λ. If this is allowed, equations (5) can be derived directly from a Hamiltonian function H for the total system particle-field:

$$H = \frac{1}{8\pi}\sum_\lambda \sigma_\lambda^2(a_\lambda^* a_\lambda + b_\lambda^* b_\lambda) + H^{(\mathrm{mat})}. \tag{6}$$

Here, $(\mathrm{i}/4\pi)(\sigma_\lambda/c)^{1/2}a_\lambda^*$, $(1/4\pi)(\sigma_\lambda/c)^{1/2}a_\lambda$, as well as $(\mathrm{i}/4\pi)(\sigma_\lambda/c)b_\lambda$, $(1/4\pi)(\sigma_\lambda/c)b_\lambda^*$ are to be regarded as canonically conjugate pairs of variables. The first expression on the right in (6) is precisely the transverse portion of the electromagnetic energy of the total field:

$$\frac{1}{8\pi}\sum_{\lambda}\sigma_{\lambda}^{2}(a_{\lambda}^{*}a_{\lambda}+b_{\lambda}^{*}b_{\lambda})=\frac{1}{8\pi}\int(\boldsymbol{E}_{\text{tr.}}^{2}+\boldsymbol{H}^{2})\,\mathrm{d}V. \tag{7}$$

Quite apart from convergence difficulties, (6) has objections associated with it, because the transverse part of the electromagnetic mass is now counted twice, in any case if the m in $H^{(\text{mat})}$ is supposed to be the experimental mass.

The representation of the theory for which I am responsible finally also arrives at a formalism that is equivalent to the use of (6) as a foundation, i.e. precisely to a formalism which, after quantization, leads to the usual quantum theory of radiation. However, the interpretation is different. As the *proper field* of the particle, I introduce the difference between the total field and the external field. For the transverse portion of the proper field, in which we are interested first of all, this yields:

$$\vec{\boldsymbol{F}}_{\text{tr.}}^{\text{eigen}}=\vec{\boldsymbol{F}}_{\text{tr.}}^{\text{total}}-\vec{\boldsymbol{F}}_{\text{tr.}}^{\text{ext}}. \tag{8}$$

We now assume that the proper field may be approximately represented by that field which corresponds to a uniformly moving particle which, at the time under consideration, would always have the position and velocity of the actually existing particle (dragged-along Coulomb field). In the neighborhood of the particle, this proper field has precisely the right singularity of which we spoke above, and its energy corresponds precisely to the transverse electromagnetic portion of what is customarily called the kinetic energy of the particle (complemented by the rest energy). Furthermore, this field precisely corresponds to what one would call the difference between the total field and the external field in the case of quasi-stationary motion. Of course, with this definition, the external field in the neighborhood of the particle will generally still grow strongly and specifically like r^{-1}; the vector potential $\vec{A}^{\text{ext}}$ of the external field, which appears in $H^{(\text{mat})}$, remains finite, however.

Let us designate the Fourier components of the transverse proper field by a''_{λ}, b''_{λ}:

$$a''_{\lambda}=a_{\lambda}-a'_{\lambda},\ b''_{\lambda}=b_{\lambda}-b'_{\lambda} \tag{9}$$

and (5) can be brought to the form

$$\dot{a}'_{\lambda}+2\pi\mathrm{i}\nu_{\lambda}a'_{\lambda}=-\mathrm{i}\frac{16\pi^{2}c}{\sigma_{\lambda}}\frac{\partial H^{(\text{mat})}}{\partial a'^{*}_{\lambda}}-(\dot{a}''_{\lambda}+2\pi\mathrm{i}\nu_{\lambda}a''_{\lambda})=0 \tag{10}$$

(similarly for b'_{λ}), or

$$\frac{\mathrm{d}}{\mathrm{d}t}(a'_{\lambda}\exp 2\pi\mathrm{i}\nu_{\lambda}t)=-\mathrm{i}\frac{16\pi^{2}c}{\sigma_{\lambda}}\frac{\partial H^{(\text{mat})}}{\partial a'^{*}_{\lambda}}\exp 2\pi\mathrm{i}\nu_{\lambda}t-\frac{\mathrm{d}}{\mathrm{d}t}(a''_{\lambda}\exp 2\pi\mathrm{i}\nu_{\lambda}t) \tag{11}$$

(similarly for b'_{λ}).

Now it does not appear possible to bring these equations, together with the equations of motion for the particle, to canonical form without neglecting something. The appearance of terms with $\dot{\vec{v}}$, which determine the radiation field (there contained in the last term of (11) because the a''_{λ} depend on the velocity), already is a sign of this. Now, there are many problems, however, where one is interested in the first place only in the manner in which the radiation field is built up over times that are long compared with the characteristic periods of the radiating particle. The radiation of an electron bound in the field of an atom is an example of this, likewise the radiation from an electron flying past a nucleus. In the latter case, the collision time would be the period characteristic for the problem. So that the radiant energy becomes arbitrarily large in the course of time, one can imagine that all of space is uniformly filled with electrons flying in the same direction. In such cases, the Fourier components of the proper field will now fluctuate back and forth between finite limits and the time average of the expression $(\mathrm{d}/\mathrm{d}t)(a''_\lambda \exp 2\pi \mathrm{i}\nu_\lambda t)$ is therefore equal to zero. In such a case, therefore, the *secular change of the external field* is presumably correctly described by

$$\frac{\mathrm{d}}{\mathrm{d}t}(a'_\lambda \exp 2\pi \mathrm{i}\nu_\lambda t) = -\mathrm{i}\frac{16\pi^2 c}{\sigma_\lambda}\frac{\partial H^{(\mathrm{mat})}}{\partial a'^*_2} \exp 2\pi \mathrm{i}\nu_\lambda t. \tag{12}$$

Giving up the correct rapidly periodic tremblings of the electromagnetic field, the change of the particle-field system could thus be described by the Hamiltonian function:

$$H = \frac{1}{8\pi}\sum_\lambda \sigma_\lambda^2(a'^*_\lambda a'_\lambda + b'^*_\lambda b'_\lambda) + H^{(\mathrm{mat})}. \tag{13}$$

The only difference in (13) compared to (6) now lies in the interpretation of the coefficients a_λ, b_λ; from the present perspective, these would be the Fourier components of the external field. We thus can interpret H as the sum of the energy of the external field and the energy $H^{(\mathrm{mat})}$, which we would associate with the particle in the case of quasi-stationary motion; the latter implicitly contain the energy of the proper field, because the kinetic energy of the particle appears in $H^{(\mathrm{mat})}$, expressed by means of the experimental mass constant. In the previous interpretation of H, certain energy contributions of the total system were counted twice, so to speak.[3] On the other hand, in the present interpretation, one must say that a certain portion of the total energy has been omitted, and specifically that energy which, in the complete expression for $(1/8\pi)\int(\boldsymbol{E}^2 + \boldsymbol{H}^2)\,\mathrm{d}V$ corresponds to the inner product of the external field and the proper field. One could designate it as the interference energy of these two fields.

Equation (13) now automatically also contains a reaction of the external field on the particle; according to the present interpretation, this is described correctly only in a secular sense. Even before the secularization process, we did indeed give up a precise analysis of the reaction of the field on the particle; but one can prove that, in a non-relativistic approximation, the laws of this reaction as they are known from electron theory are implicitly contained in the canonical laws of motion, if these laws are combined with (10).

If *several particles* are present, and if we define the proper field as the sum of the dragged-along Coulomb fields, it should be noted that, for each particle, the external field corresponds only to a part of what we would call the external field in the case of ordinary quasi-stationary motions. The proper fields of the other particles also act as an external field at the position of the particle under consideration. But it is well known that, in a not strictly relativistic approximation, this mutual action of the proper fields can be taken into account by adding certain terms. This primarily involves the Coulomb potential $\sum e_k e_l/r_{kl}$, but also the so-called Darwin terms, which describe those effects of the magnetic and electric interaction that are proportional to $1/c^2$. In the quoted paper, I have proven why inclusion of these terms in $H^{(\mathrm{mat})}$ precisely corresponds to considering the 'interference energy' of the proper fields. If one now proceeds to the quantization of (13), the above-mentioned interaction of the proper fields (Coulomb forces, etc.) has already been prepared in the form of $H^{(\mathrm{mat})}$ as a consequence of our point of

[3] During the discussion, this difference was especially underscored by Mr Pauli.

view. In contrast to usual quantum dynamics, it does not appear as a consequence of a quantized field theory. With the unsatisfactory situation of the entire theory, I do not believe that this circumstance presents a strong argument against our conception. The formalism, according to which the exchange of energy with the radiation field is treated, naturally does not differ in any way from the usual one. We regard the entire description as allowable only in an approximate sense, when there are reasons to consider the secularization process as meaningful. This circumstance, however, first of all causes us to lend credence only to results of perturbation calculations of first order (emission and absorption of radiation and the reactions coupled therewith) and specifically in those cases where the probability that an elementary process takes place grows by a very small amount within the characteristic period of the system. However, the recent papers by Bloch and Nordsieck and by Pauli[4] show that it must be possible to extend the usual formalism of the quantum theory of radiation in such a fashion that neither will any difficulties stand in the way of a consistent treatment of the 'multiple processes' which – in terms of the correspondence principle – are of a simple type.

[4]Compare Mr Pauli's lecture at this conference.

Index to Frame-setting essay*

*Throughout, I have abbreviated quantum electrodynamics as QED.